SpringerBriefs in Energy

T0172132

More information about this series at http://www.springer.com/series/8903

Manfred Hafner • Simone Tagliapietra •
Giacomo Falchetta • Giovanni Occhiali

Renewables for Energy Access and Sustainable Development in East Africa

 Springer

OPEN

Manfred Hafner
Future Energy Program
Fondazione Eni Enrico Mattei
Milano, Italy

Simone Tagliapietra
Future Energy Program
Fondazione Eni Enrico Mattei
Milano, Italy

Giacomo Falchetta
Future Energy Program
Fondazione Eni Enrico Mattei
Milano, Italy

Giovanni Occhiali
Future Energy Program
Fondazione Eni Enrico Mattei
Milano, Italy

ISSN 2191-5520   ISSN 2191-5539   (electronic)
SpringerBriefs in Energy
ISBN 978-3-030-11734-4   ISBN 978-3-030-11735-1   (eBook)
https://doi.org/10.1007/978-3-030-11735-1

Library of Congress Control Number: 2019931914

This book is an open access publication.

This Springer imprint is published by the registered company Springer Nature Switzerland AG.
The registered company address is: Gewerbestrasse 11, 6330 Cham, Switzerland

Foreword

Imagine life with no electricity. No fridge with vaccines for a newborn child, nor important anaesthetics for the mother giving birth. No exposure to computers in school. No machines to produce. And yet more than a billion people live without.

Then consider the difficulty of the analyst wishing to study how to electrify a country and supply energy to drive its development. They are faced with a mountain of a task. Data is limited; modelling tools are expensive.

Ponder the plight of the policymaker who should manage national resources, set market rules and engage with investors and development partners. The latter want hard auditable information. In many instances that type of information simply does not exist. Investment is therefore not made. Those without electricity stay without and, in turn, are locked out of the opportunities that most of the readers of this book will enjoy.

In this book, the authors join the charge to help breach this deadlock—with a focus on East Africa. They do so by mapping the high potentials of renewable and gas reserves in the region, as well as the complementary role they may play. Then, by taking advantage of scientific advances in open data and free open models, they deliver quantitative scenarios. These provide a powerful vision. Together with thought-leading analysis the authors help orientate the debate around—and potential directions for—investment and development in electrification, gas and renewable energy deployment.

Along the way they leave as a legacy transparent, reproducible, reconstructable data, tools and insight. This is a unique and powerful contribution.

Division of Energy Systems Analysis (KTH-dESA)   Mark Howells
Royal Institute of Technology
Stockholm, Sweden

Acknowledgements

Support from the Fondazione Eni Enrico Mattei (FEEM) in realising this book is gratefully acknowledged. The authors are also thankful to the KTH Royal Institute of Technology—Division of Energy System Analysis (dESA) for organising the 2018 *Summer School on Modelling Tools for Sustainable Development*.

About the Fondazione Eni Enrico Mattei (FEEM)

The Fondazione Eni Enrico Mattei (FEEM), founded in 1989, is a non-profit, policy-oriented, international research centre and a think tank producing high-quality, innovative, interdisciplinary and scientifically sound research on sustainable development. It contributes to the quality of decision-making in public and private spheres through analytical studies, policy advice, scientific dissemination and high-level education. Thanks to its international network, FEEM integrates its research and dissemination activities with those of the best academic institutions and think tanks around the world.

About FEEM's *Future Energy Program* (FEP)

The *Future Energy Program* (previously called *Energy Scenarios and Policy* program) aims to carry out interdisciplinary, scientifically sound, prospective and policy-oriented applied research, targeted at political and business decision-makers. This aim is achieved through an integrated quantitative and qualitative analysis of energy scenarios and policies. This innovative and interdisciplinary approach puts together the major factors driving the change in global energy dynamics (i.e. technological, economic, geopolitical, institutional and sociological aspects). FEP applies this methodology to a wide range of issues (energy demand and supply, infrastructures, financing, market analyses, socio-economic impacts of energy policies).

Contents

About the Authors

Manfred Hafner is the Coordinator of the *Future Energy Program* at FEEM. He is Professor of International Energy Studies, teaching at the Johns Hopkins University School of Advanced International Studies (SAIS Europe) and at the SciencesPo Paris School of International Affairs (PSIA). He also teaches in many Executive Education master's and MBA courses worldwide. He has more than 30 years of experience in consulting on international energy issues for governments, international organisations and industry. He is/was a member of several high-level intergovernmental cooperation networks and councils. He has a long track record of interdisciplinary research coordination. He holds several master's degrees: in engineering from the Technical University of Munich; in economics and business from the IFP School, the University of Paris-2 Pantheon-Assas and the University of Bourgogne; and in energy policy and management from the University of Pennsylvania. He obtained his PhD in Energy Studies with "summa cum laude" at Mines ParisTech (Ecole des Mines de Paris).

Simone Tagliapietra is Senior Researcher at the *Future Energy Program* of the Fondazione Eni Enrico Mattei. He is Adjunct Professor of Global Energy Fundamentals at the Johns Hopkins University School of Advanced International Studies (SAIS Europe) and Research Fellow at Bruegel, the European economic think tank. He is also Lecturer at the Università Cattolica del Sacro Cuore, Senior Associate Research Fellow at the Istituto per gli Studi di Politica Internazionale and Non-resident Fellow at the Payne Institute of the Colorado School of Mines. He is an expert in international energy and climate issues, with a record of numerous publications covering the international energy markets, the European energy and climate policy and the Euro-Mediterranean energy relations. He obtained his PhD in International Relations at the Università Cattolica del Sacro Cuore in Milan.

Giacomo Falchetta is a Researcher at the *Future Energy Program* of the Fondazione Eni Enrico Mattei, where he carries out applied research on the dynamics of access to electricity and on the power–climate nexus in Sub-Saharan Africa, with a particular focus on the use of remotely sensed data. He is also a PhD candidate at the Università Cattolica del Sacro Cuore, and he has been a Guest Research Scholar at the International Institute for Applied Systems Analysis (IIASA). He holds an MSc in Environmental Economics and Climate Change from the London School of Economics.

Giovanni Occhiali is an Overseas Development Institute Fellow at the National Revenue Authority of Sierra Leone. Previously, he worked as a Researcher at the *Future Energy Program* of the Fondazione Eni Enrico Mattei. He holds a PhD in Economics from the University of Birmingham, an MSc in Development Economics from SOAS University of London and a BSc in Economics and Political Science from the University of Bologna.

Abbreviations

AfDB	African Development Bank
BCA	Benefit-Cost Analysis
Bcm	Billion cubic metres
CF	Capacity Factor
CIA	Central Intelligence Agency
CO	Carbon Monoxide
CO_2	Carbon Dioxide
CSP	Concentrated Solar Power
EA	East Africa = EA-8: Burundi, Kenya, Malawi, Mozambique, Rwanda, South Africa, Tanzania, Uganda
EA-7	East Africa (EA-8) excluding South Africa
EA-8	East Africa including South Africa
EAC	East African Community
EAPP	Eastern Africa Power Pool
FDI	Foreign Direct Investment
FiT	Feed-in-Tariff
GIS	Geographical Information System
GDP	Gross Domestic Product
GNI	Gross National Income
GW	Gigawatt
GWh	Gigawatt hour
HFO	Heavy Fuel Oil
HVDC	High Voltage Direct Current
IEA	International Energy Agency
IHME	Institute for Health Metrics and Evaluation
IPCC	Intergovernmental Panel on Climate Change
IEA	International Energy Agency
IPP	Independent Power Producer
IRENA	International Renewable Energy Agency
kV	Kilovolt

kWh	Kilowatt hour
LCOE	Levelised Cost of Electricity
LPG	Liquefied Petroleum Gas
MBTU	Million British Thermal Units
MG	Mini-Grid
Mt	Megaton
MW	Megawatt
NG	Natural Gas
ODA	Official Development Assistance
OECD	Organisation for Economic Co-operation and Development
OnSSET	Open-Source Spatial Electrification Tool
PJ	Petajoule
PPA	Power Purchase Agreement
PPP	Purchasing Power Parity
PV	Photovoltaic
RCT	Randomised Control Trial
RE	Renewable Energy
RE-FiT	Renewable Energy Feed-in-Tariff
SDG	Sustainable Development Goal
SA	Standalone system
SPPA	Standardised Power Purchase Agreement
SSA	Sub-Saharan Africa
ST	Short ton
TWh	Terawatt hour
UN	United Nations
UNDP	United Nations Development Programme
USD	United States Dollar
US EIA	United States Energy Information Administration
V	Volt
VRE	Variable renewable energy
WTP	Willingness-to-pay

Chapter 1
Introduction

Contents

East Africa (EA) is a weakly-defined macro-region, with its extent varying from a geographical, cultural, and political perspective depending on the context inquired. This book makes the explicit choice of referring to EA as the area including Burundi, Kenya, Malawi, Mozambique, Rwanda, Tanzania, and Uganda (Fig. 1.1). Furthermore, despite not strictly belonging to the region, South Africa is also accounted for in the analysis.[1] This is due to the strong ties and interdependencies in the energy and economic sectors that South Africa exhibits with EA-7 countries, and to the lessons that can be learned from its emblematic case—an outlier in terms of energy and economic development.

As of 2018, the population of EA-8 stands at 271 million (Table 1.1). On average, it has grown at an annual rate of 2.6% over the last 5 years (World Bank 2017) and it is projected to keep an increasing trajectory. As a result, under a medium fertility scenario it would reach 569 million units by 2050 (United Nations Population Division 2017). EA-7 is also one of the fastest growing regions in the world. In 2017, the regional real GDP grew by an estimated 5.9%, although with substantial country heterogeneity, and it is forecasted to keep a similar pace in the coming years (AfDB 2018). Notably, the industrial sector has been growing at a double pace *vis-à-vis* agriculture, dragged by a skyrocketing mining activity. The demand-side is playing a considerable role in pushing economic growth, with consumption and public investment in infrastructure paving the way. Nonetheless, the aforementioned growth has hitherto led to only limited poverty reduction and the real GDP/capita in EA-7 countries is still among the world's lowest. Many inhabitants of the region face extreme poverty, lack access to safe water and health facilities and exhibit high malnutrition and low education levels, while inequality indicators are also stagnating.

---

[1]Throughout the book, the EA-8 acronym denotes all EA countries including South Africa, while EA-7 excludes it.

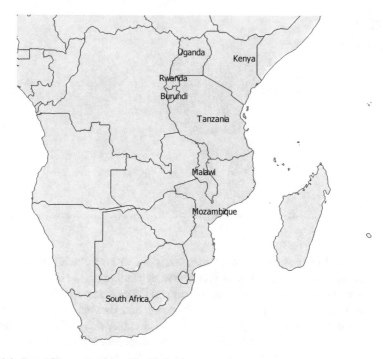

Fig. 1.1 East Africa as considered in this book

The key challenge faced by EA-7 countries is therefore to promote their national development, with the emblematic objective of abandoning their status of low or lower middle-income economies. Those are defined as countries *"having a per capita gross national income of US $1006 to $3955"* (World Bank). As of today, South Africa is the only member of that club from the region. In this context, one of the main obstacles (if not the main one) for unleashing the regional growth-potential is to develop its energy resources and its power sector and provide access to sustainable energy to the entire population. Currently, EA-8 hosts 3.6% of the global population but accounts for only 1.5% of total global primary energy consumption (IEA 2017a), which becomes 0.2% if South Africa is excluded. Moreover, as shown in Fig. 1.2, the regional population served by some form of access to electricity currently stands at 39% (IEA 2017b), with very low per-capita consumption levels and more than 150 million people living without access.

It is important to outline that the successful achievement of most development objectives is highly dependent on meeting the challenge of electrification and improved energy access, due to the strong interdependencies between the energy sector and virtually all other economic activities. Since, in principle, some development targets could show a certain degree of competition rather than complementarity, especially in low-income countries where available resources are scarce, action needs to be taken with a multi-level approach, i.e. it needs to consider impacts and externalities on the economic, social and environmental spheres. With these issues in

Table 1.1 Socio-economic figures for countries in East Africa

	Burundi	Kenya	Malawi	Mozambique	Rwanda	Tanzania	Uganda	South Africa
Population (million)	10.5	48.5	18.09	28.83	11.92	55.57	41.49	55.91
Population growth rate (2015–16) (%)	+3.1	+2.6	+2.9	+2.9	+2.4	+3.1	+3.3	+1.6
Urban population (%)	12	26	16	33	30	32	16	65
Rural population (%)	88	74	84	67	70	68	84	35
PPP GDP (mil. Int. $)	8187	153,000	21,155	35,089	22,803	150,336	76,702	739,419
PPP GDP/capita (int. $)	778	3156	1169	1217	1913	2787	1849	13,225
IHDI[a]	0.276	0.391	0.328	0.280	0.339	0.396	0.341	0.435

Sources: UN (2015), World Bank (2017), and UNDP (2016)

[a]Inequality-adjusted human development index, where a higher value indicates a combination of greater average achievements in health, education and income, and lower inequality

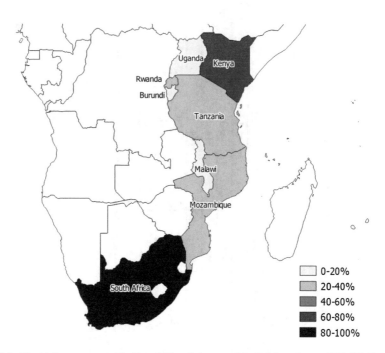

Fig. 1.2 Electricity access rates in East Africa. Source: authors' elaboration on IEA (2017a)

the foreground, in 2015 the General Assembly of the United Nations agreed to work towards 17 *Sustainable Development Goals* to be achieved by 2030, with their progress being measured through 169 targets (United Nations 2015). The framework is one of multi-objective development, where economic growth is to be achieved alongside affordable and clean energy diffusion, poverty eradication, zero hunger, good health, and other development goals. As far as this book is concerned, the achievement of one of these goals is particularly relevant, i.e. the SDG 7, which sets the objective of *'ensuring access to affordable, reliable, sustainable and modern energy for all'*, with focus on access to electricity and to clean cooking; on increasing the share of renewable energy (RE) in the total generation mix; and on the upgrade of energy services in developing countries.

A multitude of studies have been carried out to assess the potential of renewable energy (RE) in enabling the development of the energy sector, the expansion of electricity access and thus the development of countries in Sub-Saharan Africa (SSA). Here, we report some of the most recent and influential contributions from international institutions and academic scholars, to later introduce the elements of innovation brought by our analysis.

IRENA's *Africa 2030: Roadmap for a Renewable Energy Future* (2015) provided a comprehensive roadmap for Africa's energy transition. It focused on RE, and it concluded that half of the development potential from modern renewable energy resources and technologies would be represented by biomass-based heat applications, which would progressively displace traditional biomass combustion. The report:

(1) builds on a regional assessment of supply, demand, renewable energy potential, and technology prospects; (2) discusses country-by-country the role of enabling policies and of regulatory framework in catalysing investment, as well as other measures to attract investors and to promote off-grid renewable solutions so to increase energy access and boost welfare. PBL Netherlands Environmental Assessment Agency's *Towards universal electricity access in Sub-Saharan Africa* (2017) estimated the investment requirements for achieving universal electricity access by 2030 (as put forward by SDG 7) at additional USD 9–33 billion per year in the 2010–2030 period on top of business-as-usual investment. In the report, the key role of decentralised electrification systems is outlined, both in terms of mini-grids and stand-alone systems. Furthermore, it claims that while renewable energy generation will become increasingly competitive, fossil fuels would continue to play a significant role in future electricity production in SSA. However, the report also suggests that achieving SDG7's target would have only a small impact on global CO_2 emissions, compared to a situation in which this target is not achieved. The IEA's 2017 *Energy Access Outlook* (2017b) agreed upon the fact that decentralised systems, and mainly solar PV, would be the least-cost solution for roughly 75% of the additional connections needed in SSA. The report then focused on the issue of clean cooking, and it stressed the necessity of a transition from solid biomass through the deployment of LPG, NG and electricity in urban areas, and a range of technologies (improved biomass and cooking stoves) in rural areas. Furthermore, the involvement of local communities, especially women, when designing solutions was highlighted.

On the specific case of expanding and improving energy access in EA, Othieno and Awange (2016) authored a chapter on energy resources in EA in their book *Energy Resources in Africa*. They observed that only a small amount of locally available energy resources has been developed in the region. The main underlying problems that the authors identified are associated with the low level of local technological capacity in RE technologies and the hitherto inadequate support for energy development initiatives. Furthermore, throughout their book they highlighted that information on specific energy sites that could guide potential investors, including their commercial viability, is not readily available.[2] The fact that energy provision policies in EA countries are not well coordinated and hence very difficult to implement is also evidenced. Moreover, the authors argued that most RE resources have not been properly recognized for their commercial significance at the national level, and thus that an imbalance in the level of support to the development of different energy resources has been witnessed in the past years. Byakola et al. (2009) produced a report on *Sustainable Energy Solutions in East Africa* based on the experiences and policy recommendations from NGOs in Tanzania, Kenya and Uganda. They pinpointed the effectiveness of investment in small and medium-scale renewable energy projects among the rural and urban poor. Conversely, they

---

[2]Recent open-data and open-source modelling initiatives from the World Bank and SolarGIS, the IRENA, and the KTH Royal Institute of Technology are precisely aimed at tackling the challenge.

listed the key challenges faced by international actors involved in the electrification process in EA. The issues include the fact that national policies and institutional framework are not giving sufficient leverage for entrepreneurs to consolidate or tap into new energy business ventures, and thus that potential entrepreneurs face high initial investment costs and associated risks. On the demand-side, they highlighted the fact that most potential end-users of modern energy services cannot afford to pay upfront for the products and services offered by the entrepreneurs while end-user financing is equally not well-instituted. Overall, they underlined the need to foster participation of different stakeholders in decision-making. They suggested a standardised roadmap of energy project development with the following steps: (1) identification and selection of target areas, (2) local information collection, (3) participatory problems identification, (4) needs and opportunities assessment, (5) demonstration and awareness raising, (6) capacity-building for scaling-up through market development, (7) participatory monitoring and evaluation, and (8) learning and replication with adaptation. REN21 and UNIDO's (2016) *Renewable Energy and Energy Efficiency* report on EA highlighted that the East African Community is the second largest single regional market in Africa and one of the fastest growing regions in SSA. It reviewed the key features of the energy sector, and provided a renewable energy market and industry overview, discussing heat, electricity, and transport uses. It then highlighted targets, power support policies, and it discussed the question of energy efficiency, with consideration on the energy intensity with which the system develops, as well as on transmission and distribution networks and on policy aimed at favouring efficiency in use. Concerning investments, the report stated that between 2009 and 2013, the East African Community (EAC) attracted approximately USD 5.8 billion in aid from the international community, with more than 80% of total EAC energy investments having been channelled into geothermal and wind projects in Kenya. Kammen et al. (2015) demonstrated how the combination between a growing demand for energy and a series of new fossil fuel discoveries in the region, coupled with a better understanding of its RE potential, have led to a defining moment for the uptake of a sustainable (or unsustainable) regional resource management trajectory and development path. They highlighted the significance of the Eastern African Power Pool (EAPP) in the process as a driver of investment, a catalyser for the on-grid and off-grid investments, and a mechanism to mitigate climate (e.g. droughts and derived hydropower disruptions) and energy demand-related (e.g. peaks) risks. They also discussed the role of Kenya as an emerging clean energy leader in EA and in SSA, with visions of 5000 MW of new on-grid capacity in only 40 months. Hansen et al. (2015) discussed solar PV policies and diffusion in EA with a regional perspective, individuating two emerging trends: (1) a movement from donor and government-based support to market-driven diffusion of solar PV; and (2) a transition from small-scale, off-grid systems towards mini-grids or large-scale, grid-connected solar power plants. The authors identified three key drivers behind the ongoing transitions, namely the decline in the cost of PV units and thus of the levelized cost of PV, the role of international donors, and domestic policy frameworks such as feed-in-tariffs. Interestingly, the paper also discussed some of the reasons which are likely to have paved the way to Kenya becoming the RE leader in the region. Among these, the growing middle-class, the favourable geographic conditions (in terms of cheap wind and geothermal potential), local subcomponent suppliers and backers, and the existence of a national business culture.

Overall, these and further studies highlighted the following key points for energy development in EA-7: (1) potential, both in RE and hydrocarbons is large and more than technically enough to guarantee energy self-sufficiency, (2) the development of the integrated power pool network is of upmost importance to minimise risks and cut costs, (3) enabling conditions play a big role, and Kenya is being a virtuous example in this sense, (4) the EA-8 region is currently in a critical juncture for its energy development trajectory: local policymakers should be supported by researchers and the international community with resource mapping, least-cost electrification scenarios, and policy recommendations based on previous success stories.

This book contributes to this literature and it presents different elements of innovation with respect to the previous studies. It specifically focuses on EA, accounting for the local configuration of access and resource endowment, and addressing economic, technical and policy questions. In Chaps. 2 and 3, it presents a standardised and extensive energy resource mapping, highlighting the *status quo*, development plans, ongoing energy infrastructure projects—including grid expansions—enacted regulation, and overall untapped technical potential. The focus is put on RE sources for power generation purposes (i.e. excluding the direct combustion of solid biomass), including solar PV and CSP, wind, hydropower, and geothermal. A comprehensive RE atlas (including maps, estimated potential on a resource-basis, and current policy in place) for the specific case of EA could in fact not be found among previous recent publications. In Chap. 4, least-cost electrification scenarios are modelled to provide policy-relevant insights on which level of penetration of different technologies would be required to achieve 100% electrification by the year 2030 while also satisfying the growing demand from already electrified household and the emerging industrial sector. The chapter also covers both the capacity additions required and the total investments necessary to achieve this objective. The analysis is not limited to a modelling exercise, since in Chap. 5 specific attention is paid to the main issues faced in the accomplishment of a faster, more inclusive and cost-effective energy access in EA-7. This part covers technological, economic, cooperation, policy, and financing conditions, as well as the opportunities and risks involved in the development of a portfolio of renewables to promote energy security in a sustainable way. Chapter 6 then discusses the challenges and opportunities that might stem from the interaction between local RE potential and NG resources currently under development in the region, while also referring to the results of the electrification modelling exercise. To conclude, policy recommendations based on our results and targeted at international cooperation and development institutions, local policymakers, and private stakeholders in the region are elaborated.

References

AfDB (2018) East Africa economic outlook 2018. African Development Bank. https://www.afdb.org/en/documents/document/east-africa-economic-outlook-2018-100840/. Accessed 3 Aug 2018

Byakola T, Lema O, Kristjansdottir T, Lineikro J (2009) Sustainable energy solutions in East Africa–Status, experiences and policy recommendations from NGOs in Tanzania, Kenya and Uganda. Friends of the Earth Organization, Norway

Hansen EU, Harmsen M, Nygaard I (2015) Review of solar PV policies, interventions and diffusion in East Africa. https://www.researchgate.net/publication/273790226_Review_of_solar_PV_policies_interventions_and_diffusion_in_East_Africa. Accessed 23 Apr 2018

IEA (2017a) World energy outlook 2017

IEA (2017b) WEO 2017 special report: energy access outlook. International Energy Agency

IRENA (2015) Africa 2030: roadmap for a renewable energy future. http://www.irena.org/publications/2015/Oct/Africa-2030-Roadmap-for-a-Renewable-Energy-Future

Kammen DM, Jacome V, Avila N (2015) A clean energy vision for East Africa: planning for sustainability, reducing climate risks and increasing energy access. https://rael.berkeley.edu/wp-content/uploads/2015/03/Kammen-et-al-A-Clean-Energy-Vision-for-the-EAPP.pdf. Accessed 23 Apr 2018

Othieno H, Awange J (2016) Energy resources in East Africa. In: Energy resources in Africa. Springer, Switzerland, pp 33–137

PBL Netherlands Environmental Assessment Agency (2017) Towards universal electricity access in Sub-Saharan Africa. PBL Netherlands Environmental Assessment Agency. http://www.pbl.nl/en/publications/towards-universal-electricity-access-in-sub-saharan-africa. Accessed 23 Apr 2018

REN21, UNIDO (2016) EAC renewable energy and energy efficiency status report 2016. http://www.eacreee.org/document/eac-renewable-energy-and-energy-efficiency-status-report-2016. Accessed 23 Apr 2018

The World Bank (2017) World Bank Data.. Accessed 20 Nov 2017

UNDP (2016) Human development report 2016. UNDP. http://www.undp.org/content/undp/en/home/librarypage/hdr/2016-human-development-report.html

United Nations (2015) Resolution adopted by the general assembly on 25 September 2015: transforming our world: the 2030 agenda for sustainable development

United Nations Population Division (2017) World population prospects: the 2017 revision

Chapter 2
East Africa: Regional Energy Outlook

Contents

EA-8 hosts 3.6% of world's population but only accounts for 1.5% of total global primary energy consumption (The World Bank 2017; IEA 2017a). With South Africa (the second economy of SSA and an outlier in the region) excluded, this unbalance gets even more pronounced, with 2.9% of world's population consuming 0.2% of total primary energy. While the share of EA-8 population without access to electricity has fallen from 90% in 2000 to 61% in 2016 (IEA 2017b), the absolute number of people without access has instead increased by eight million as electrification efforts have been outpaced by rapid population growth.

Electricity consumption in the region stood at 261 TWh in 2015, with that of South Africa alone at 227 TWh, and the remaining countries having consumed only 34 TWh (IEA 2017a; CIA 2017). Just to provide a comparison, in the same year a high-income country such as Italy consumed 310 TWh of electricity, despite having less than a fifth of EA-8's population (CIA 2017). EA-8's energy demand is primarily concentrated in the residential sector (Fig. 2.1), with solid biomass employed for cooking and lighting purposes (on average accounting for 65% of primary energy consumption in the region). This situation outlines a vast potential of energy supply expansion in other sectors such as industry, transport and tertiary. Expanding energy access across EA would bring an array of both direct and indirect benefits, including the improvement of socio-economic conditions (employment, firm profitability), as well as the enhancement of health, agricultural productivity and water access.

The poor state of both the electricity grid and of transboundary interconnections in the region, as well as that of other energy infrastructure, represents another critical problem. Existing infrastructure predominantly serves urban centres hosting the minority of the population, while its majority, dispersed over large rural areas, has to satisfy its energy need largely through solid biomass (Fig. 2.2).

Biomass includes wood fuels, agricultural by-products, and dung, usually collected at the household level or traded in informal markets, and it is employed for the

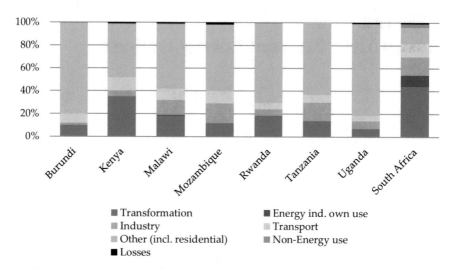

Fig. 2.1 Primary energy demand by sector in EA countries. Source: Authors' elaboration on IEA (2017a) and UN Energy Statistics Yearbook (2015)

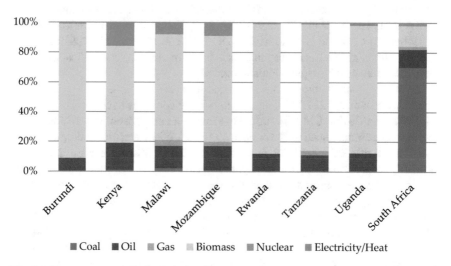

Fig. 2.2 Primary energy mix in EA countries. Source: Authors' elaboration on IEA (2017a) and UN Energy Statistics Yearbook (2015)

bulk of cooking, lighting, and heating activities. The only exception to this trend in the region is represented by South Africa, where coal dominates the energy mix. A range of adverse socio-economic and health effects result from the consumption of traditional biomass. Such impacts have been extensively documented in different spheres (see Bandyopadhyay et al. 2011), with particular attention on the resulting indoor air pollution. Other issues connected with biomass consumption range from productivity losses due to the burdensome and time-demanding fuel collection

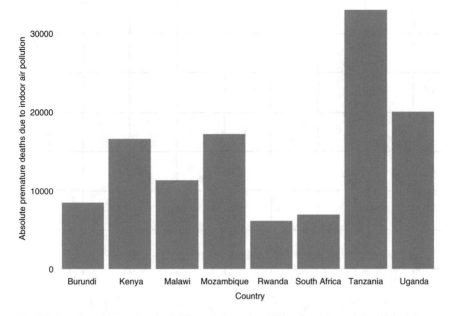

Fig. 2.3 Number of premature deaths due to indoor air pollution from the use of solid fuels in year 2016. Source: Authors' elaboration on IHME (2017)

process (especially among women and children, refer to Biran et al. 2004) to health concerns at both collection and consumption phases[1]; from increasing rates of environmental degradation to the very low energy efficiency resulting from households' rudimentary equipment. There is robust evidence that indoor air pollution has led to around 120,000 death in EA-8 countries in 2016 alone (IHME 2017). Figure 2.3 provides an overview of the magnitude of the mortality burden in each EA country.

At the same time, access to electricity across EA-8 currently averages 35% (or 28% without South Africa), with 150 million people without access and notable rural-urban inequality in most countries (Fig. 2.4).

Hydropower is the dominant source in the electricity generation mix of Burundi, Kenya, Malawi, Mozambique, Tanzania, and Uganda (Fig. 2.5), with Mozambique and Tanzania still having large untapped potential. Overall, total installed hydropower capacity in EA-7 stands at 3 GW. South Africa leads by far in terms of coal generation, with an overall coal-fired capacity of 40 GW.

Given the current situation, a well-managed and long-sighted energy development and electrification plan throughout EA-8 would not only represents a step forward in achieving greater energy access (SDG 7), but also accelerate the

---

[1]The earlier because of the dangers of injuries, the latter through the breathing of combustion gases, chiefly CO, particulate matter, and biogenic volatile organic compounds, as well as burning injuries.

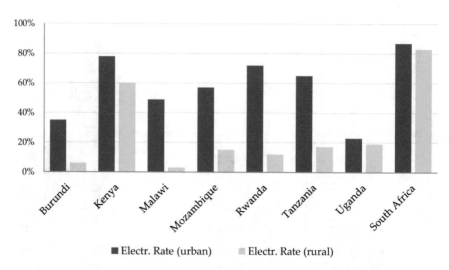

Fig. 2.4 Electricity access rates in urban (left) and rural (right) in EA. Source: Authors' elaboration on IEA (2017b)

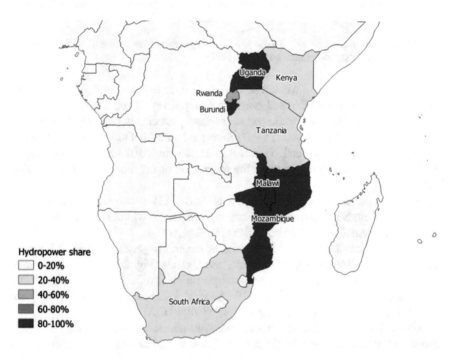

Fig. 2.5 Hydro share (%) over total generation in EA. Source: Authors' elaboration on US EIA (2017) and Hydropower and Dams (2017)

realization of other development targets. Multiple international institutions (UN, IEA, World Bank) have agreed that initiating a transition to modern energy on the scale implied by the ambitious electrification targets set by EA countries (Table 2.1) will require the development of a portfolio of diverse technologies, energy sources, and grid expansion solutions, each specific to its own context.

As the next chapters will show in detail, EA-8 countries are endowed with substantial energy resources and technical generation potential (Table 2.2): solar PV potential is abundant and widespread throughout the region, and overall it stands at 219,500 TWh/year.[2] CSP (gross 176,000 TWh/year in the region) is mostly feasible in South Africa and Kenya. Untapped hydropower, both at large and small scale, is found to varying degrees in all countries except South Africa. The same is true for geothermal, in particular in the northern part of the Rift Valley area shared between Kenya, Uganda and Rwanda and further down south in Malawi. Wind potential stands at 16,600 TWh/year (considering areas with a wind turbine capacity factor greater than 40%). Bioenergy for generation purposes is a further viable option, mostly in South Africa and Kenya. Hydrocarbon resources are also abundant, but their distribution is highly skewed: Uganda is the only country with substantial oil reserves (standing at 2.5 billion barrels), as those of Kenya and South Africa are less prominent or less accessible; the NG endowments of Mozambique and Tanzania are large (together they sum to 4200 bcm), while those of Rwanda and South Africa are only partially viable. Finally, South Africa is the richest country in terms of coal resources (with reserves standing at 35 billion tons), while mining activity is also taking place in Mozambique (which could have more than 20 billion tons of reserves), Malawi, and Tanzania.

Irrespective of energy resources endowment and RE potential, as of today all EA-7 countries (i.e. with the exception of South Africa) share a structural lack of secure and universal energy access, while regional energy development processes face several barriers connected to various technological, economic, financial, institutional, political, and social issues. All these dimensions will be discussed throughout this book. For instance, variable renewable energy (VRE) sources such as solar PV and wind power require firming, which entails either interlinked networks, flexible backup supplies, storage and or demand response. Large-scale storage of electricity is technologically possible but expensive and challenging at grid level scale. As VRE potential is brought on-line, additional investments will hence be required to ensure high degrees of grid and supply reliability. Nevertheless, hitherto national and local governments have lacked an effective governance (and the financial means) to give the onset to a process of diffused electrification exploiting domestic potential. Greater public-private and local-international coordination are required to ensure effective policy support for exploiting this untapped capacity in order to bring more RE on-line.

---

[2]The following RE potential figures are drawn from (IRENA 2014), and they consider gross technical potential in all suitable areas (i.e. they do not account for economic viability). Hence, the figures are interesting in comparative rather than in absolute terms.

Table 2.1 The power sector in EA countries: current figures and targets

Country	Burundi	Kenya	Malawi	Mozambique	Rwanda	Tanzania	Uganda	South Africa
Installed capacity (MW)	41	2269	353	2687	211	1583	947	46,963
Electrification rate (2017) (%)	10	65	11	29	30	33	19	86
Urban (%)	35	78	49	57	72	65	23	87
Rural (%)	6	60	3	15	12	17	19	83
El. Rate target (year)	25% (2025)	100% (2020)	30% (2020)	100% (2025)	70% (2018)	50% (2025)	26% (2022)	100% (2020)

Sources: IEA (2017b), The World Bank (2017) and CIA (2017), various governmental energy strategies and visions

Table 2.2 Power capacity, generation, RE untapped potential, and hydrocarbon reserves (by country)

Country	Burundi	Kenya	Malawi	Mozambique	Rwanda	Tanzania	Uganda	South Africa
Installed capacity (MW)	41	2269	353	2682	175	1583	947	46,963
Generation[a] (GWh)	263	9139	2098	17,740	476	6219	3856	252,578
Hydro potential[b] (TWh/year)	5	25	47	96	–	32	14	74
PV potential (TWh/year)	888	23,000	5200	22,000	900	39,000	9500	42,200
CSP potential (TWh/year)	786	15,400	4500	16,800	800	31,500	8600	43,300
Wind potential (TWh/year, 40% CF)	–	1.800	43	5	–	790	24	1560
Geothermal potential (MW)[c]	18	5000–10,000	4000	147	20–300	650	450	–
Biomass generation potential (PJ)[d]	–	260	–	950	–	530	–	1160
Natural gas reserves (bcm)[e]	0	0	0	2830	55–60	1614	14	0
Proven coal reserves (Mt)	0	0	2.20	1975	0	297	0	35,400
Oil reserves (billion barrels)	0	0.77	0	0	0	0	4.5	0

[a]CIA (2017)
[b]Zhou et al. (2015). NB: the figure represents exploitable (technically and economically feasible) hydropower potential
[c]Omenda and Teklemariam (2010)
[d]Deng et al. (2015), referring to all lignocellulosic crops (no food)
[e]Hydrocarbon resource endowment figures come from BP (2017) and ENI (2017a, b)

Particularly in rural areas, roadblocks of different kinds are faced. Not only there exist heavy financial and geographical barriers to the extension of the national grid, various issues are also encountered when communities and households face the decision of whether to invest in appliances and local grids such as solar home systems or small hydro networks. In such cases, the available incentives and subsidies, payment methods or schemes, combined with collective behaviour and social norms, are the defining factors for appliance adoption. Thus, there is a need to ensure effective communication across stakeholders to clearly understand the implications of policy support to deploy private capital in electrification projects. Digital technologies can play an important role in this sense. While private companies are starting to gain a relevant market share in EA-8, with 140 million USD raised in the region in 2015 alone, more than half of the global total (REN21 and UNIDO 2016), there is still a long way to go to achieve energy security. International investors, including China, the World Bank and the African Development Bank, are playing a significant role in the process of addition of large capacity projects and grid extension. However, the effectiveness of their investment is often not maximised because of country specific institutional and policy factors. Thus, there exists a tight link between local policy and politics, financial support and international cooperation.

In this context, the next chapters provide a more detailed insight into the specific energy situation in each country, as well as into their RE potential, resource endowment, and policy frameworks in place. This background is then employed to produce least-cost full electrification scenarios and model them within an electrification tool (OnSSET), to later discuss the key technical, economic, and policy challenges faced by policymakers in the exploitation of untapped RE potential in the region.

References

Bandyopadhyay S, Shyamsundar P, Baccini A (2011) Forests, biomass use and poverty in Malawi. Ecol Econ 70:2461–2471. https://doi.org/10.1016/j.ecolecon.2011.08.003
Biran A, Abbot J, Mace R (2004) Families and firewood: a comparative analysis of the costs and benefits of children in firewood collection and use in two rural communities in Sub-Saharan Africa. Hum Ecol 32:1–25. https://doi.org/10.1023/B:HUEC.0000015210.89170.4e
BP (2017) Statistical review of World Energy 2017. BP
CIA (2017) The world factbook 2017
Deng YY, Koper M, Haigh M, Dornburg V (2015) Country-level assessment of long-term global bioenergy potential. Biomass Bioenergy 74:253–267. https://doi.org/10.1016/j.biombioe.2014.12.003
ENI (2017a) Volume 2—World gas and renewables review 2017.. https://www.eni.com:443/en_IT/company/fuel-cafe/world-gas-e-renewables-review-2017.page. Accessed 15 Jan 2018
ENI (2017b) Volume 1—World oil review 2017.. https://www.eni.com:443/en_IT/company/fuel-cafe/world-oil-gas-review-eng.page. Accessed 15 Jan 2018
IEA (2017a) World energy outlook 2017
IEA (2017b) WEO 2017 special report: energy access outlook. International Energy Agency

Institute for Health Metrics and Evaluation (IHME) (2017) Global burden of disease from household air pollution

IRENA (2014) Estimating the renewable energy potential in Africa: a GIS-based approach. http://www.irena.org/publications/2014/Aug/Estimating-the-Renewable-Energy-Potential-in-Africa-A-GIS-based-approach. Accessed 23 Apr 2018

Omenda P, Teklemariam M (2010) Overview of geothermal resource utilization in the East African rift system

REN21, UNIDO (2016) EAC renewable energy and energy efficiency status report 2016. http://www.eacreee.org/document/eac-renewable-energy-and-energy-efficiency-status-report-2016. Accessed 23 Apr 2018

The International Journal on Hydropower and Dams (2017) Hydropower and dams in Africa 2017. https://www.hydropower-dams.com/product/africa-map-2017/

The World Bank (2017) World Bank Data.. Accessed 20 Nov 2017

UN Energy Statistics Yearbook (2015) Energy statistics yearbook 2014. United Nations Publications, New York

US EIA (2017) International energy statistics. https://www.eia.gov/beta/international/data/browser/#/?c=4100000002000060000000000000g000200000000000000000001&vs=INTL.44-1-AFRCQBTU.A&vo=0&v=H&end=2015. Accessed 23 Apr 2018

Zhou Y, Hejazi M, Smith S et al (2015) A comprehensive view of global potential for hydro-generated electricity. Energy Environ Sci 8:2622–2633. https://doi.org/10.1039/C5EE00888C

Chapter 3
Country-Level Analysis: Power Sector, Energy Resources, and Policy Context

Contents

3.1   Burundi

3.1.1   Electricity Access, Installed Capacity, and Non-renewable Reserves

With an installed capacity of 41 MW and a total generation of 300 GWh in 2015 (United Nations 2015), Burundi is characterised by particularly low electricity access figures. Only 10% of the ten million inhabitants is served by electricity, with the rate reaching 35% in urban areas and dropping to 6% in rural areas (IEA 2017b). Households located in the capital Bujumbura account for the bulk of electricity consumption. Electricity represents however just 1.3% of the national energy consumption, and—as seen in Fig. 3.1—most of the generation capacity (91%) comes from hydropower, with two small (<50 MW) plants active in Lake Kivu, with the other main source of generation being represented by diesel plants (RISE 2017).

The national utility REGIDESO (Control and Regulation Agency for the Water and Electricity) benefits from a long-term legally established monopoly for electricity supply, transmission, and distribution. The distribution system is modest in bearing and extent, with 546 km of transmission grid and 337 km of distribution lines in place. This represents a further constraint to new capacity additions. Burundi has no fossil fuel endowments, and the country has been often struggling to import oil products to operate generators and plants.

3.1.2   RE Potential

Burundi has very large untapped potential for hydropower development (with a technical potential of 1700 MW, of which 300 MW are seen as commercially viable) with four projects equating to 90 MW currently being developed and four others being planned (Hydropower and Dams 2017). This potential could be highly beneficial to the filling of the growing supply gap in the Bujumbura capital area,

Fig. 3.1 Burundi's electricity generation mix. Source: Authors' elaboration on US EIA

■ Diesel   ■ Small Hydro

9%

91%

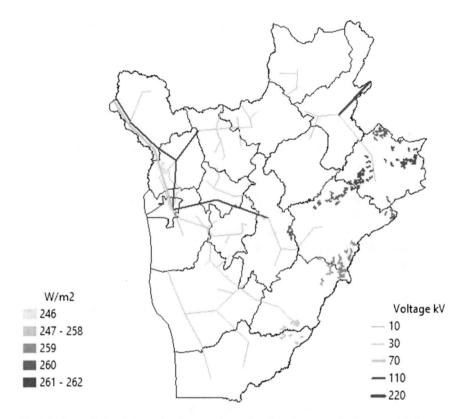

Fig. 3.2 Burundi electricity grid and most relevant locations in terms of solar potential. Source: Authors' elaboration on IRENA—REmap (2017)

where demand for electricity is soaring. Furthermore, 18 MW of geothermal potential have been identified, and there are also areas with wind and solar potential suitable for exploitation (Fig. 3.2). Regarding the latter, as of January 2018 a 7.5 MW PV plant with a purchasing power agreement (PPA) already in place is under construction. Some use of PV for lightning and public buildings has also been reported.

3.1.3   RE Policy Framework

On the policy side, the *Decentralized Rural Electrification Strategy (2015–2017)* aims to maximize the social impact of decentralized RE to bring the benefits of modern energy technologies to rural children and families and promote transfer of skills and approaches to institutional, commercial, and community level structures. Furthermore, the medium-term *Vision Burundi 2025,* approved in 2011 by the

UNDP and the Government of Burundi, aims to achieve an electrification rate of 25% by 2025, reducing wood burning for heating and cooking in households while focusing on micro and mini renewable plants (including hydro). The country has indeed a fairly developed legal and policy framework for mini-grids (updated in 2015), including ownership and operation by private companies and the presence of duty exemptions for PV array and modules and for power generators. "Solar Electricity service with Mini Grids in Africa-Burundi" (SESMA-Burundi) as recently submitted a project aiming to bring online the first 7 mini grids of the country, currently at the feasibility study stage. To accomplish the electrification rate target set in *Vision Burundi 2025*, the Government has adopted a Decentralized Rural Electrification Strategy in 2015 and it plans to establish a National Agency for Renewable Energy and Energy Efficiency. The year 2011 saw the establishment of both the Burundian Agency for Rural Electrification (ABER) and of REGIDESO, which functions as the controller of water and electricity supply as well as of implementation, monitoring and application of tariffs. In compliance with law I/014 of year 2000, the public services of water and electricity provision are liberalized and regulated. That is, while the energy sector remains a public service under the responsibility of the state, its doors are as open to public as they are to private investors, selected through invitation to tender with specific criteria.

3.2   Kenya

3.2.1   Electricity Access, Installed Capacity, and Non-renewable Reserves

In Kenya around half of the installed capacity and of total generation stems from hydropower (Fig. 3.3).

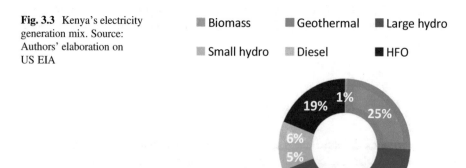

Fig. 3.3 Kenya's electricity generation mix. Source: Authors' elaboration on US EIA

Total generation capacity sums to 2269 MW, which produced 9139 GWh in 2015 (CIA 2017; Kenyan Energy Regulatory Commission). According to UNdata, hydro-power is followed by thermal generation units (heavy fuel oil and diesel-fired) for installed capacity and electricity generated, while following ERC data this seems true only for installed capacity, given the higher reported relevance of electricity generated by geothermal units (which also account for around a fourth of total capacity in the country). There are also 25 MW of wind and some biomass plants in place, although their contribution to electricity generation is still minimal. The industrial sector is currently the main consumer of electricity (around 57%), followed by domestic customers (26.3%) and by commercial and public services (refer to tables in the *Appendix*).

According to IEA (2017a), electricity access in Kenya stands at 65% (78% in urban areas and 60% in rural areas). Since 2012, when the electrification rate was around 20%, the number of connected users has more than tripled as a result of government initiatives such as the *Last Mile Connectivity Project*, of private-public partnerships and international support and to the successful appearances of pay-as-you-go solar home system companies and innovative business models for mini-grid development. Thanks to this positive developments, Kenya is currently on track to reach universal electricity access by 2030 and it is identified as one of the virtuous examples in Sub-Saharan Africa in terms of electrification objectives (Gordon 2018). Still, more than half of the country's households are not yet connected to the national grid, and increasing efforts are required on the supply-side to satisfy a steeply increasing demand for power in cities.

Concerning fossil fuel endowments, in 2016 oil reserves amounting to 766 million barrels (and as much as 1.63 billion barrels of gross oil contingent resources) have been discovered in Lokichar, in the North of the country (Africa Oil Corp. 2016). The government plans to embark on large-scale oil production and build a pipeline to connect the fields with harbours on the East coast. The oil would predominantly be shipped to Asia. A heated debate is taking place on the set-up of the distribution mechanism of anticipated revenues.

3.2.2 RE Potential

Significant undeveloped hydropower resources exist, including 1449 MW of large hydro and 3000 MW of small hydro, across 260 sites with good potential in areas with high population density and energy demand (Fig. 3.4). Feasibility studies for 12 sites with a combined capacity of 33 MW were carried out in 2013, while other 14 are currently ongoing. Additionally, Kenya has around 20 feasible sites for the development of geothermal (which currently accounts for 27% of installed capacity), with a combined capacity between 5000 MW and 10,000 MW (the government-owned Geothermal Development Company is tasked with the integrated develop-ment of geothermal planning). The country also has a good potential for solar generation (see Fig. 3.5: Kenya receives solar irradiation of 4–6 kWh/m^2/day) and

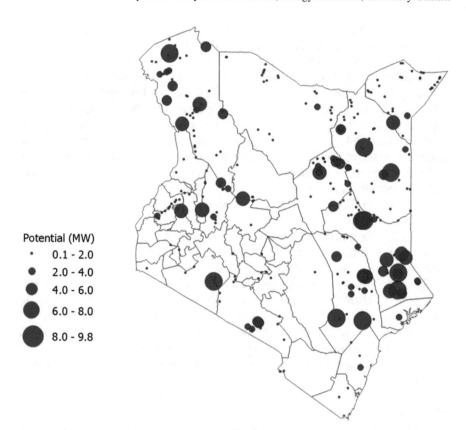

Fig. 3.4 Small hydropower potential in Kenya. Source: Authors' elaboration on Korkovelos et al. (2018)

a thriving market for small PV (12–50 Wp), with a governmental plan for providing electricity to educational, health, administrative and private sites far from the grid. 977 institutions have already been reached through such program, for an installed capacity of more than 1.5 MW at peak time; in addition, isolated diesel stations are being transformed into diesel-solar hybrid with four other similar systems being currently deployed in rural areas. There are also a handful of large solar projects, each of around 40 MW, which received PPAs in late 2015. The country is also characterised by a high wind potential of up to 346 W/m^2 (with average wind speed exceeding 6 m/s in many parts of the country), with one wind farm of 25.5 MW being operated by KenGen and another one, Lake Turkana wind farm, whose 310 MW operated by an independent power producer (IPP) should come on-line in late 2018 (Fig. 3.5) also figure as currently unexploited potential. The 19% planned contribution of nuclear energy is also noticeable, with a first 1000 MW plant expected to be operational by 2027 (currently in feasibility study phase).

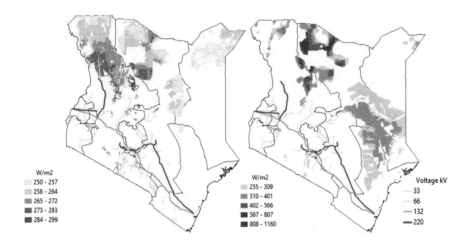

Fig. 3.5 Kenya electricity grid and most relevant solar and wind potential. Source: Authors' elaboration on IRENA—REmap (2017)

3.2.3 RE Policy Framework

The generation and transmission system of Kenya is based on a rolling 20-year least cost plan (Least Cost Power Development), which was updated in 2011 to reflect the *Vision 2030* objectives. Such plan targets the contribution of geothermal energy to grow to 20% of installed capacity and the share of hydropower—currently the most relevant source of electricity in the country—to decrease to 5%, while wind power should increase its share in the generation portfolio from less than 1–9%.

The 2006 *Energy Act* set the objective of promoting the development of all RE sources, charging the Ministry of Energy with the creation of both an Energy Regulatory Commission—responsible of production, distribution, supply and use of RE—and of a national research agenda on RE. Since then, Kenya has also been one of the few EA-8 countries with tiered RE feed-in-tariffs (introduced in 2008, to be later revised in 2010 and 2012), i.e. differentiated by large-scale and small-scale projects (with 10 MW as the threshold) and generation sources. The regulatory framework introduced tariffs at which IPPs are authorised to sell electricity at a fixed price for a fixed term of 20 years. Tariffs reflect the generation cost and they should not exceed the long-term marginal cost for on-grid systems.

Concerning recent energy policy initiatives, two key plans have been set forth: the Rural Electrification Master Plan and the Distribution Master Plan. The former sets the objective of achieving a 65% access to electricity by 2022 and full access by 2030. Clear steps and supporting instruments are defined. The Rural Electrification Authority is the main authority responsible for tracking progress of the plan. The latter, put forward in 2013, produced estimates of the long-term annual investment required in all distribution infrastructure, from 66 kV to LV, up to 2030. To serve this purpose, the Kenya Energy Bill was passed in 2015 in order to establish a distribution licensee plan and to put into place the requisite electric supply lines necessary to enable any person in the licensee's supply area to receive electrical energy either directly from the licensee or from an accordingly authorised electricity retailer. Furthermore, in the coming years Kenya intends to also roll out an auction-based development plan for wind and solar and to replace the ongoing feed-in tariff scheme, as well as introducing net metering for residential generation and to establish regulations for mini-grids (Climatescope 2017).

3.3   Malawi

3.3.1   Electricity Access, Installed Capacity, and Non-renewable Reserves

UNdata reports an installed capacity of 501 MW for the year 2014, while the publicly-owned electricity supply company of Malawi ESCOM (2017) and CIA (2017) refer to an installed capacity of 353 MW. The discrepancy between the figures likely stems from 210 MW of self-producing diesel units not attached to the grid, and from rehabilitation work on certain power stations (e.g. undergoing dredging operation at the pond reservoir at Nkula Power Station). Note that the vast majority (>95%) of grid-available capacity in Malawi comes from hydropower (Fig. 3.6), and due to rain-failure in 2015 and 2016 there has been a significant reduction in the hydro-based generation from ESCOM, which has effectively averaged around 200 MW. Further reduction in the available hydro capacity is expected due to continuously dropping water level in Lake Malawi as a result of both increased withdrawals and reduced runoff.

Overall, in 2015 electricity generation stood at 2120 GWh (CIA 2017), with roughly 11% of the population being served by the grid. The proportion of urban residents with access to power is a magnitude larger than that of rural residents (49% and 3% respectively, IEA 2017a). Similar shares of electricity are being consumed

Fig. 3.6 Malawi's electricity generation mix. Source: Authors' elaboration on US EIA

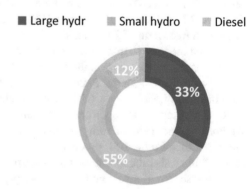

by the manufacturing and household services, which represent the bulk of total consumption.

With regards to fossil fuels, Malawi is endowed with 20 million tons of proven coal reserves (estimated resources are between 80 million and one billion tons), at four coal fields. Although up to now coal was mostly used for industrial heating, a 300 MW coal plant to generate electricity is currently under development, with project planning at an advanced stage and commissioning expected by 2021. Coal would be imported from Mozambique by rail, given the plant's location close to the border between the two countries and the already existing nearby railway line. On the other hand, Malawi is not endowed with either oil or gas reserves, although it is locked in a border dispute with Tanzania over Lake Nyasa, which might contain both. No exploration will take place until the dispute is resolved.

3.3.2 RE Potential

Following *Malawi Growth and Development Strategy III (2017–2020)*, environmental and social impact assessment studies for further hydropower projects (three stations of 350 MW, 200 MW and 120 MW respectively) are currently under development. Malawi has also some solar (Fig. 3.7) and wind (Fig. 3.8) potential (which is currently undergoing a mapping project, refer to World Bank 2017), and although non-hydro renewables currently contribute very modestly to power generation, their share should increase in the near future with the development of 21 IPPs solar Schemes (3 of which already possess a PPA) that should increase the available grid capacity to 563 MW. However, as of today only two IPPs will have firm PPAs in place: HE Power for a 41 MW hydro project, and IntraEnergy for a 120 MW coal plant (Climatescope 2017). Malawi is not connected to power systems with neighbouring countries and therefore cannot benefit from its membership in international energy cooperation plans. A plan to link the national grid with Mozambique and Zambia for the purchase of at least 150 MW of electricity is under development (Climatescope 2017).

3.3.3 RE Policy Framework

In Malawi, the Malawi Rural Electrification Program (MAREP) was last updated in 2017 and is currently undergoing its eight phase since its inception in 1980. It is supported by the Malawi Energy Regulatory Authority (MERA) and Rural Electrification Management Committee. Phase 8 foresees connecting to the grid 336 new trading centres by the end of 2018, along with generation capacity additions. To attract foreign investment in the power sector, the country has also started an opening of the electricity market with a standardized PPA for IPPs to operate. In 2017, ESCOM also held its first RE tender, contracting 70 MW of solar PV at four

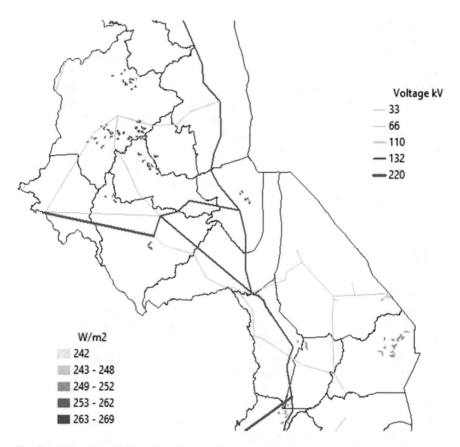

Fig. 3.7 Malawi electricity grid and most relevant locations in terms of solar potential, zoom. Source: Authors' elaboration on IRENA—REmap (2017)

sites. Furthermore, the existing *Rural Electrification Programme* uses revenue from a 3% levy on electricity consumption and fuel import taxes to fund the extension of the national grid to business centres. The country aims to increase energy access to 30% by 2020 and 40% by 2050. The National Energy Policy, still to be enforced, seeks to further diversify the energy mix with a major focus on renewable sources, such as solar and wind. The new energy policy is expected to raise the RE target to 22% by 2030. Moreover, all the public institutions including hospitals and schools should gain access to electricity by 2035 through grid connection or mini-grid and off-grid projects (Climatescope 2017). Currently, those with access rely on isolated diesel generators. Furthermore, in 2012 the Malawi Energy Regulatory Authority (MERA) drafted a feed-in tariff plan including small scale hydro, PV, biomass, wind, and geothermal. For instance, for the case of hydro the tariffs apply for 20 years from the date of the first commissioning of the plant and they range between 0.08 and 0.14 USD/kWh depending on the project's scale and on the nature of the investor (firms or individuals), while for PV generation the feed-in-tariff (FiT) stands

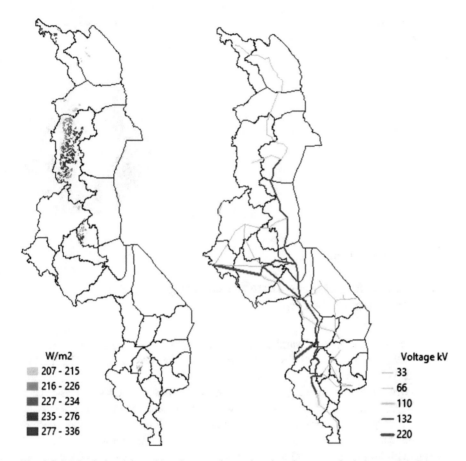

Fig. 3.8 Malawi electricity grid and most relevant locations in terms of wind potential. Source: Authors' elaboration on IRENA—REmap (2017)

at 0.20 USD/kWh, for biomass and geothermal at 0.10 USD/kWh, and for wind at 0.13 USD/kWh. The policy also states that the FiTs policy shall be subject to review every 5 years from the date of publication. Any changes that may be made during such reviews shall only apply to RESE power plants that shall be developed after the revised guidelines are published.

Fig. 3.9 Mozambique's electricity generation mix. Source: Authors' elaboration on US EIA

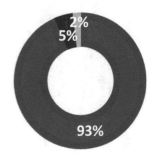

3.4   Mozambique

3.4.1   *Electricity Access, Installed Capacity, and Non-renewable Reserves*

The installed capacity in Mozambique is of 2687 MW (96% is hydropower, as seen in Fig. 3.9), while generated electricity stands at 17,739 GWh (CIA 2017), the first among EA-7 countries.

However, the vast majority of the capacity is located at Cahora Bassa Dam (2075 MW), owned (92.5%) by Electricidade do Moçambique (EdM) but serving predominantly South Africa (1575 MW) with two 533 kV high voltage direct current lines connected to the plant. On the other hand, there is no interconnection between the plant and the southern part of Mozambique.

There is a discrepancy in thermal power generation data, standing at 359 MW according to UNdata and only at 65 MW for the World Bank. The difference might be partially due to a diverse way of considering the 175 MW gas plant in course of development with SASOL. Differences are noted in the reported data about electricity consumption by sector. Overall, it seems that the share of electricity accruing to residential customers is just slightly smaller than that going to the different kind of industrial users.

Mozambique has significant fossil fuel endowments. Gas reserves are estimated at 2830 bcm (ENI 2017a, b) and are pushing a strong development of the resource (prone to render Mozambique a regional leader), while coal reserves of good quality coking coal are estimated at 20 billion tons. The size and quality of reserves have led to the decision of developing a few coal-fired power stations: a 300 MW plant in the north of the country (of property of a coal mine owner, currently finalising the financing phase), and two others—300 MW near Moatize and 150 MW nearby Chirodzi—mostly to serve another coal mine, while excess electricity will be sold to the grid. However, the financing for the latter two plats is proving to be an issue.

3.4.2   RE Potential

Mozambique has considerable unexploited electricity potential. The existing capacity of 2.1 GW of hydropower is just over 10% of its potential (19 GW), with a further 1 GW of small hydro potential (Fig. 3.10). 351 hydro-projects for a total of 5.6 GW are identified as high priority for development. Most of these projects (236) are below 5 MW. Although geothermal potential is still under investigation, at least 147 MW of technically feasible capacity have been identified, of which 20 have been marked as a priority. Furthermore, of the 2.7 GW of grid-connectable solar generation potential, only 599 MW are currently considered for development due to the present limitations imposed by short circuit grids in place. Furthermore, solar PV also has a good potential for the off-grid electrification of rural areas (Fig. 3.11). So far it is estimated that 2.25 MW of PV have been installed in rural areas, while the market potential for off-grid pico-PV is estimated at 75 MW, with a further 4.6 MW of solar-diesel hybrid. Finally, the overall wind potential is estimated at 4.5 GW, with 1.1 GW viable for connection, of which 230 MW are considered a priority (Fig. 3.12). To exploit potential at a large-scale, Mozambique faces the great challenge of implementing improved electricity transmission and distribution systems.

3.4.3   RE Policy Framework

In 2009 the *Policy on the Development of New and Renewable Energy* explicitly promoted the use of RE resources for meeting the development needs of Mozambique, with a particular focus on increasing the access to modern energy in rural areas. The *Policy* outlined the framework of incentives for their development and suggested the creation of an investment priority plan. The *Policy* begun to be operationalised in 2011 through the *Strategy for the New and Renewable Development 2011–2025*, which divided actions between those directed to on- and off-grid development, with a focus on large scale PV programmes for lightning and water pumping and heating. It provided import tax exemption for RE equipment, VAT exemption for rural electrification and expansion projects, and corporate tax exemption for companies investing in either of the latter. Moreover, in 2011 the *Public-Private Partnership* law was also published, opening a space for IPPs, which however must sell electricity directly to EDM (the national electricity company) and negotiate prices on a contract-by-contract basis. As a result, in 2018 CRONIMET Mining Power Solutions and MOSTE have signed a MoU with FUNAE (the Rural Electrification Agency) to develop Mozambique's first privately developed and financed mini-grid (expected to generate up to 200 kWp of solar power) on Chiloane Island, which will also be the largest pre-paid solar mini-grid in the country. Upon successful implementation of the Chiloane Island mini-grid, the consortium expects to develop a portfolio of 60 or more mini-grids across Mozambique. Further on the policy side, Decree 58/2014

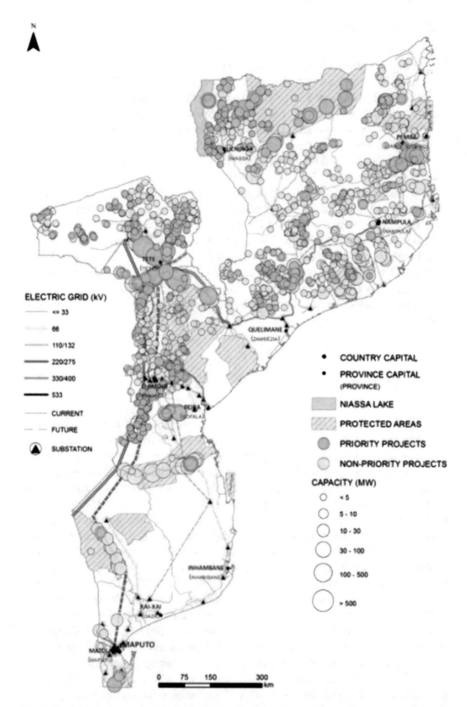

Fig. 3.10 Small hydropower sites examined by priority and potential, and the national electricity grid. Source: Renewable Energy Atlas Mozambique (2014)

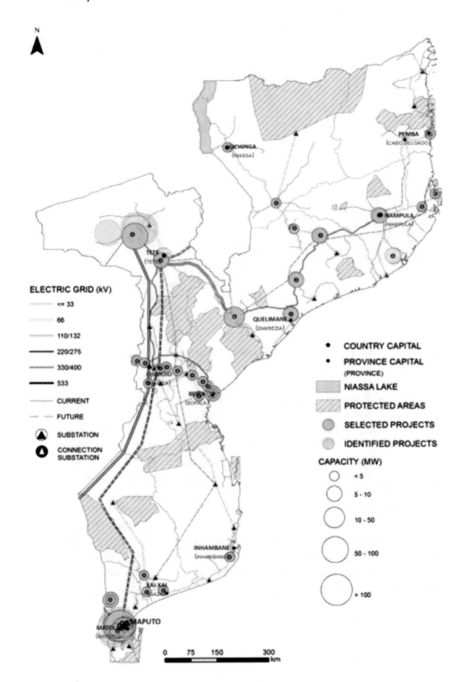

Fig. 3.11 Solar potential sites examined by status and potential, and the national electricity grid.
Source: Renewable Energy Atlas Mozambique (2014)

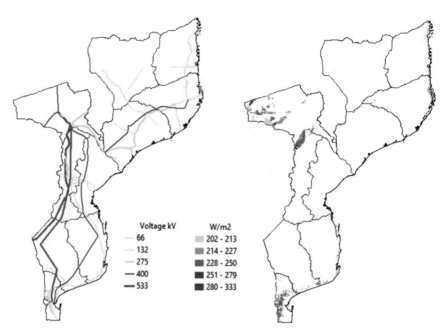

Fig. 3.12 Mozambique electricity grid and most relevant wind potential. Source: Authors' elaboration on IRENA—REmap (2017)

created Mozambique's feed-in tariff, which applies to biomass, wind, small hydro and solar projects from 10 kW to 10 MW. Prices vary according to technology and capacity (ranging between 0.07 USD/kWh for large biomass projects up to 0.22 USD/kWh for solar PV up to 10 kW). According to this Decree, all projects must sell electricity to the state-owned utility EDM. Although the decree is available, injection of power into the grid cannot happen yet as some regulation is still to be approved. Regardless, as of late 2017, the FiT mechanism was already undergoing revision. Mozambique is also reviewing the scope of the National Electricity Council (CNELEC), the power market regulator, in order to broad and strength its role. While this might help the FiT mechanism to gain more space in Mozambique power market, there is still a long road ahead before it becomes fully in force given that the state-owned utility EDM is sanctioned as off-taker of all power contracts while already being under considerable financial strain.

Fig. 3.13 Rwanda's
electricity generation mix.
Source: Authors'
elaboration on US EIA

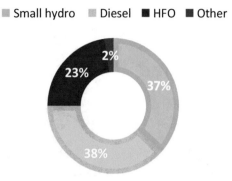

3.5 Rwanda

3.5.1 Electricity Access, Installed Capacity, and Non-renewable Reserves

In 2015 Rwanda generated most of its electricity (around 476 GWh, UNdata 2015) from hydropower (Fig. 3.13), which accounts for the bulk of installed capacity with an aggregate of 211 MW according to Rwanda Energy Group 2017. The country also has 42 MW of diesel and gas thermal plants and an 8 MW PV plant in the Eastern Province, which represents 3% of the country's on-grid generation. The figures seem to suggest that a substantial growth in total capacity was witnessed in recent years thanks to the development of gas and solar generation, although some discrepancies between REG and UN data must be noted.

The government of Rwanda has been striving to increase electricity access, which, according to IEA (2017a), stands at 30%, reaching 72% in urban areas while stopping at 12% in rural ones. According to the *Rwanda Development Board* (2017), residential access is at 40.5%, of which 29.5% comes from grid and 11% from off-grid sources. This represents a very robust improvement from access rates of just 4% in 2008 and 12% in 2012.

On the consumption side, the bulk of the electricity produced in 2014 served households (for the year 2012 the figures available through the Ministry of Infrastructure report 51% of consumption from households, 42% from medium customers, 6% from public services and 1% of exports).

Rwanda has no oil reserves, but it is endowed with 55–60 bcm of NG located under Lake Kivu. These reserves have been earmarked towards NG fired generation, with an estimated potential of 350 MW—the first 26 MW of a 100 MW project, called KivuWatt, started operating in May 2016.

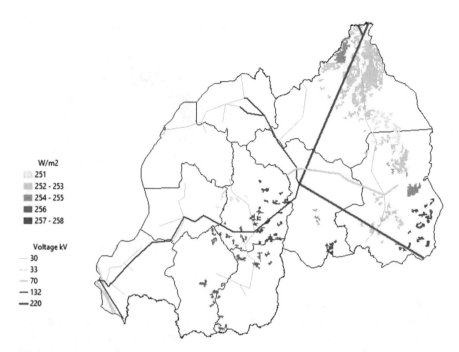

Fig. 3.14 Rwanda Electricity grid and most relevant locations in terms of solar potential. Source: Authors' elaboration on IRENA—REmap (2017)

3.5.2 RE Potential

The country has undeveloped potential in hydropower (300 MW), especially micro generation (but also inter-boundary projects with Burundi and DRC, 145 MW, at Ruzizi III, and with Burundi and Tanzania, 90 MW at Rusumo Falls) as well as a geothermal (between 170 and 340 MW, a complete study of the reserves is yet to be undertaken), peat (up to 1200 MW) and solar potential (66.8 TWh, Fig. 3.14).

3.5.3 RE Policy Framework

The overarching aims of the government is to achieve 512 MW of installed power generation capacity by 2023/24 through increases in gas, solar, hydro, peat and interconnection capacity; to increase access to electricity to 100% of the population within the same time span, with off-grid electricity reaching 48% of the total; and to decrease the reliance on biomass from 86.3% of primary energy to 50% (by 2020). To achieve such objectives, different policy instruments have been adopted. In 2009, the Ministry of Infrastructure put forward its *Electricity Access Roll Out Program*, which is being implemented by the national Rwanda Energy Group. This has been supported by 377 million USD in its first stage, which successfully increased electricity connection by 250,000 units in just 4 years. The second phase is currently ongoing, and it is being backed up by 300 million USD necessary to achieve the 70% electrification rate objective. Moreover, the European Union and the World Bank signed a 200-million-euro financing agreement in 2016 to support off-grid electrification and a $50 million agreement in 2017 to support Rwanda's Scaling Renewable Energy Program (SREP).

Over the past years, the government has also introduced an array of tools that have made on-grid development very attractive for the private sector. Since 2007, a combination of utility reforms, tenders, unsolicited proposals, a favourable tax regimes and donor support have all drawn the attention of private-sector players towards Rwanda, leading to 47 PPAs already signed to date. All IPPs must now participate in a competitive tender process, which is monitored by the Rwanda Utility Regulatory Authority (RURA). Rwanda's Renewable Energy feed-in tariff regulation was promulgated in February 2012. The Rwanda tariffs apply to small hydro from 50 kW to 10 MW. Contract terms are only 3 years, but the law specifies that the tariffs cannot be reduced. The tariffs will be reviewed in the second year of the program in order to be implemented in the third year. The Ministry of Infrastructure also subsidises 80% of new connections to the grid, providing potential customers with the ability to get a loan to cover their share of costs and to repay it to the electricity utility via a charge on electricity bills spread over 5 years. Additionally, Rwanda has signed a 30 MW PPA with Kenya, but the transmission infrastructure has not been built yet.

3.6 Tanzania

3.6.1 Electricity Access, Installed Capacity, and Non-renewable Reserves

Inconsistencies are found in the data on electric capacity in Tanzania. According to CIA (2017), in 2015 the installed capacity in Tanzania was 1583 MW, with slightly more than 50% from hydropower (Fig. 3.15). A report from the Ministry of Energy and Minerals from 2014 (*Electricity Supply Industry Reform Strategy and Roadmap 2014–2025*) states the same installed capacity but attributes only 35% of it to hydropower. Finally, the December 2016 *Power System Master Plan Update* reports an installed capacity of 1390 MW, of which 43% comes from hydropower. In any case, electricity generation in 2015 was 6025 GWh (CIA 2017).

The electricity access rate reported in the IEA's WEO (2017b) is 33%, with the usual discrepancy between urban (65%) and rural (17%) areas. Some differences between sources exists also with regards to the share of electricity consumed by different sectors. The *Appendix* reports both those for 2014 from UNdata and those for 2015 from the *Power System Master Plan Update*.

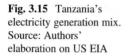

Fig. 3.15 Tanzania's
electricity generation mix.
Source: Authors'
elaboration on US EIA

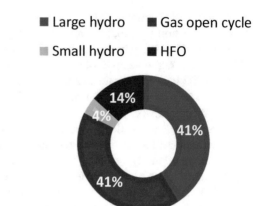

With regards to non-RE sources, Tanzania has 1600 bcm of estimated NG reserves (ENI 2017a, b) and relevant coal resources (1.9 billion tons). The country has plans to develop four gas-fired power plants, amounting to 2733 MW of capacity, and three coal-fired power plants, summing to 1400 MW.

3.6.2   RE Potential

Tanzania has significant potential for both large and small hydropower (480 MW for the latter), with SPPAs having been signed for 20.5 MW, and letter of intents for another 29.9 MW. However, similarly to neighbouring Malawi, the country has experienced extended periods of drought over the last few years, resulting in a complete switch-off of all hydropower plants in October 2015. Given that Tanzania also possess a geothermal potential of at least 650 MW, in 2013 the government created the Tanzania Geothermal Development Company in order to achieve the objectives of having 100 MW of geothermal capacity in place by 2020 and 200 MW by 2025. Tanzania would also gain from exploiting its solar energy potential, especially in its central region which receives 2800–3500 hours of sunshine per year with radiation of 4–7 kWh/m^2 per day. The potential for grid-tied PV is 800 MW, which can cover 20% of day-time peak demand (Fig. 3.16) and, so far, one SPPA has been agreed upon for a 2 MWp project in an isolated grid, for which a letter of intent signed, with several firms expressing interest in the development of 50–100 MWp solar parks. The government has removed VAT and import tax for most PV technology, reducing the end-user price, and different programs through the Rural Energy Agency are targeting rural areas through the promotion of PV adoption and the development of business models for solar companies. Furthermore, a 100 MW wind project (Singida Wind Farm) is currently under development, with at least another grid-scale generation site having been individuated (Fig. 3.17).

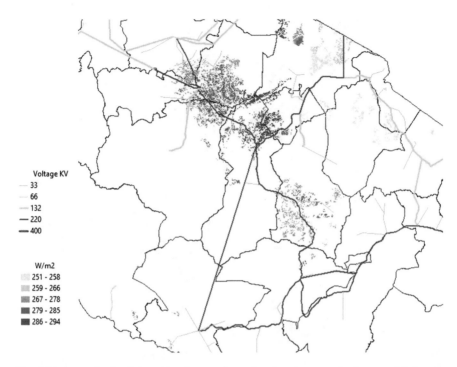

Fig. 3.16 Tanzania electricity grid and most relevant locations in terms of solar potential. Source: Authors' elaboration on IRENA—REmap (2017)

3.6.3 RE Policy Framework

On the policy side, the 2003 update to the 1992 *National Energy Policy* explicitly recognises the threat posed by climate change and calls for the promotion of the RE sector, which had so far received very little attention in the country, with capacity lacking through the whole value chain. The act mandates the establishment of an institutional and legal framework to address the technical, social and financial barriers for the diffusion of RE technologies. In 2008, feed-in tariffs were introduced to push such process but remained undifferentiated between different renewable technologies until their revision in 2015. After the revision, two different sets of prices are being applied to hydropower and biomass projects, while a bidding process is applied to solar and wind projects (although proposals developed before the revision are still subject to the old framework). Standardised PPAs are applied to all projects with capacity below 10 MW while tariffs are negotiable for larger developments. Tariffs are cost-reflective and guaranteed for the duration of the PPA (up to 25 years) but are revised annually. Distinction is made between projects feeding the national grid and those serving mini-grids, with the tariffs also changing depending on the period of the year (wet or dry season).

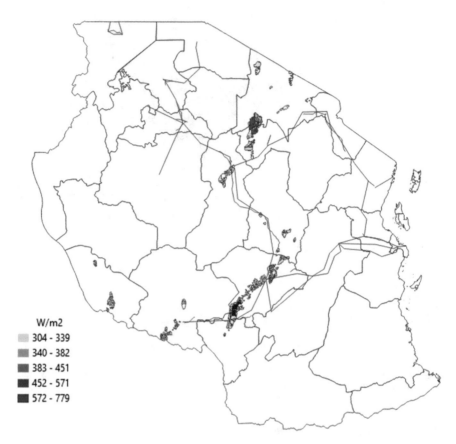

Fig. 3.17 Tanzania transmission grid and most relevant wind potential. Source: Authors' elaboration on IRENA—REmap (2017)

Moreover, Tanzania's second Five-Year Development Plan (FYDP II), presented in 2016, foresees a six-fold expansion of the power grid over the next decade. The plan sets a top-line installed base target of 10GW by 2025/26, albeit the role of renewables has not been clearly defined. Currently, under the 'optimal expansion plan', the country would add ten times more fossil fuel (predominantly goal and NG) capacity than renewables by 2030. On the other hand, Tanzania was amongst the pilot countries for the World Bank's *Scaling-up Renewable Energy Programme* (SREP) and it prepared an investment plan for RE resources in 2012. Projects financed under the scheme target rural electrification through RE deployment and the creation of mini-grids. However, issues are holding back investment, including non-payment of fees by the utility to independent generators and retail electricity rates being set too low to unlock generation opportunities upstream. At the same time, a rich network of off-grid energy providers has emerged in rural Tanzania, where the distribution of pico-solar lighting products and the development of mobile-based, pay-as-you-go business models has thrived (Climatescope 2017).

Fig. 3.18 Uganda's
electricity generation mix.
Source: Authors'
elaboration on US EIA

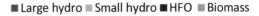

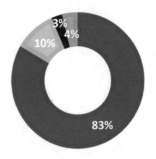

3.7 Uganda

3.7.1 Electricity Access, Installed Capacity, and Non-renewable Reserves

As of 2017, Uganda has a total installed capacity of 947 MW (Electricity Regulatory Authority), a relevant increase over the 2014 capacity of 883.3 MW (UNdata), with hydropower representing the main technology both for installed capacity (around 700 MW) and for electricity generated (around 90% of the total 3856 GWh in 2017, as shown by Fig. 3.18). A significant 18% of the total generation stems from small renewables, including small hydro, biomass co-generation (alone representing 4% of total capacity), and solar. Moreover, a recent grid capacity addition was achieved with a 10 MW grid-connected solar plant, commissioned in December 2016.

The electricity access reported by IEA (2017a) is 19%, with a connection rate of 23% in urban areas and of 19% in rural ones. The main consumer of electricity is the industrial sector (especially the iron & steel industry) consuming twice as much electricity as the commercial and domestic sectors combined.

Concerning non-renewable resources, Uganda has 2590 million barrels of oil reserves, with production set to start in 2020, and 5 bcm of gas reserves (ENI 2017a, b), most of which is associated gas.

3.7.2 RE Potential

In addition to its installed capacity, Uganda has an untapped hydropower potential of around 2000 MW (Fig. 3.19), both in large and in small hydro. With regards to the former, 2 new plants with a combined capacity of 780 MW should be online by the end of 2018 and a further 3 with a combined capacity of 1630 MW are planned by 2026. Regarding the latter, in 2015 alone 13 small hydro projects (72.6 MW) underwent feasibility studies, 6 of which (36.4 MW) were in the process of licensing

Fig. 3.19 Uganda Electricity grid and most relevant locations in terms of solar potential. Source: Authors' elaboration on IRENA—REmap (2017)

and discussion towards striking PPAs, while 11 other projects (53.8 MW) were in the pre-feasibility study stage.

Given that Uganda borders the western part of the East African Rift Valley, there is at least some geothermal potential, currently estimated at 450 MW. A prefeasibility study was undertaken for a 150 MW geothermal plant (KATWE project in Kasese), which resulted in a non-financial PPA signed by the government, a consortium of local IPPs and an American company. Geothermal exploration is addressed using the mining act, which grants a 3-year exploration licence to investors. However, the absence of a dedicated policy framework is seen as one of the main obstacles for the development of the sector.

Uganda also has a developed market for residential PV systems (6000 households and 2000 institutions have installed panels for a combined capacity of 1.1 MW as of 2014) and its first grid-level solar plant (10 MW) has been connected to the grid in December 2016 and should be serving electricity to 40,000 households located in nearby provinces. A second 10 MW plant came online in 2017 (Fig. 3.20 for solar potential). Wind power potential is only moderate within the country and is most suitable for wind mills of 2.5–10 kV capacity, which would be appropriate for small scale electricity generation or water pumping.

3.7.3   RE Policy Framework

Uganda is one of the few EA-8 countries to have liberalized its energy market: generation, transmission and supply were rendered competitive in 2001. Overall,

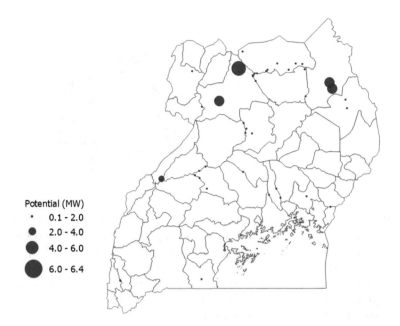

Fig. 3.20 Small hydropower potential in Uganda. Source: Authors' elaboration on Korkovelos et al. (2018)

under the *2013–22 Strategy and Plan*, the Rural Electrification Agency aims to connect over 1.4 million customers to the main grid. The Agency plans to increase today's 19% rural electrification rate to 26% by 2022, with the ultimate goal of universal access by 2035. In this direction, the 2007 *Renewable Energy Policy* was approved with the overall goal of increasing the share of RE (including hydro) in Uganda's energy consumption from 4% to 61% of the total by 2017, with a series specific capacity targets for different sources. The policy aims at increasing access to modern, affordable and reliable energy to eradicate poverty in the country. This is to be achieved through the public and private development of the large hydropower potential in the country and through the promotion of various technologies (such as mini-hydro, PV and solar water heaters) in rural and urban poor areas. Significant progress has hitherto been achieved, with more than 400 MW of large and mini hydropower capacity additions (and additional 900 MW expected to be completed by 2020), and 20 MW of solar completed between 2016–2017 and additional hundred-MW projects under development.

The *Uganda Energy Capitalisation Trust* was also created as a credit support facility to help realise the objectives of the policy, which are however currently still out of reach. The *Renewable Energy Act* also introduced feed-in tariffs for plants with a capacity below 20 MW and PPAs for RE (which were later standardised in 2014), while also setting preferential tax treatment and accelerated depreciation. The scheme was revamped in 2013 (*"Get FiT"* program) and now provides a top-up payment on the feed-in tariffs for the first 5 years of operation, an insurance against

Fig. 3.21 South Africa's electricity generation mix. Source: Authors' elaboration on US EIA

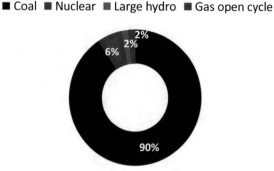

off-taker risk and simplified access to private finance. Tariffs differ by technology and the cumulative capacity limits for the tariff payment are set to increase over time. A 45% subsidy on all solar power equipment has also been in place since September 2007. Furthermore, credit-enhancement and support instruments (including technical assistance for early stage grid-scale project development and working capital for pay-as-you-go off-grid solar providers) are available to the private sector for both on- and off-grid projects (Climatescope 2017).

3.8   South Africa

3.8.1   Electricity Access, Installed Capacity, and Non-renewable Reserves

The installed capacity of South Africa is greater than that of all other Sub-Saharan African countries combined. According to ESKOM, the national utility responsible for 96% of electricity generation, in 2016 the capacity stood at 46,963 MW. Thermal generation (in particular from coal) represents the bulk of the capacity of the country (Fig. 3.21). In addition, South Africa is the only country in EA-8 which currently exploits nuclear energy (i.e. two nuclear reactors generate 5% of the country's electricity).

Generated electricity was estimated at 252,578 GWh in 2014 (UNdata), with an overall electricity access rate of roughly 86% with only minor differences between rural and urban connection rates. (IEA 2017a). The main electricity consuming sector in the country is the industrial sector, which accounts for 60% of demand—including mining and quarrying, non-ferrous metal, and chemical and petrochemical production. The household sector is responsible for the consumption of 19% of the electricity generated and commercial and public services for 14%.

Concerning fossil fuels, South Africa is one of the top-10 producers of coal worldwide and sixth for coal export, with estimated reserves of hard coal equal to 66.7 billion tons. Coal exports are anticipated to increase by 28% by 2025 due to

Fig. 3.22 South Africa electricity grid and most relevant CSP potential. Source: Authors' elaboration on IRENA—REmap (2017)

new coal plants becoming active. On the other hand, despite extensive exploration of coastal waters, only marginal gas discoveries have been made, so that gas infrastructure in the country is limited, although shale gas and coal bed methane might be present in the South Karoo Basin.

3.8.2 RE Potential

South Africa has also great unexploited endowments of RE, and the aim of the government is to achieve 9% of total electricity generation and 26% of installed capacity from renewables by 2030. Since the launch of the *Renewable Energy Independent Power Producer Programme* (REIPPP) in 2011—a competitive-bidding program backed up by an independent authority *ad-hoc* established with the aim of delivering sustainable power to the grid and creating jobs and fostering local development—more than 5 GW of renewables have been procured through four successive rounds. High potential exists for both solar and wind (Figs. 3.22 and 3.23). South Africa has one of the highest levels of solar radiation in the world, with an area of high radiation equal to 194,000 km^2. If only 1% of this area were to be developed as CSP, the generation potential would be of 64 GW. As of 2013, the target for PV and CSP for 2030 are of 9.77 GW (2 GW already procured) and of 3.3 GW (400 MW already procured) respectively. The *Integrated Energy Plan* of 2016 calls for incentives to large scale CSP with industrial steam application in the short/

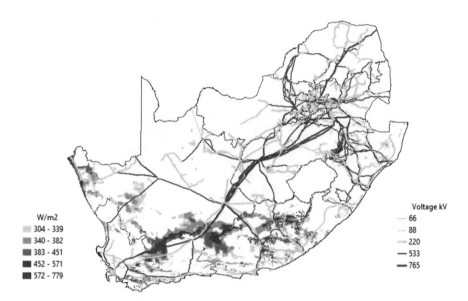

Fig. 3.23 South Africa electricity grid and most relevant wind potential. Source: Authors' elaboration on IRENA—REmap (2017)

medium term. Concerning wind potential, most of South African costal promontories have wind speeds exceeding 6 m/s corresponding to 200 W/m² of potential. The first large scale wind farm (100 MW) of Eskom started production in 2015, and of the 4.36 GW of wind-targeted by 2030, 2.67 GW have already been procured. The Department of Energy established the South African Wind Energy Programme in 2008 to provide dedicated support for wind energy development in the country, both at the industrial and R&D level, and to periodically update the South African Wind Atlas. Finally, South Africa faces various water scarcity issues, so that hydro-power does not play a particularly relevant role in its energy mix: the country currently imports 1300 MW of hydroelectricity from Mozambique, and further domestic potential is low.

3.8.3   RE Policy Framework

In November 2016, South Africa's government presented the *Integrated Resource Plan*, which outlines the country's electricity strategy to 2050. Under the plan, the country seeks to add 37GW of wind plants and 18GW of PV by 2050, while reducing the share of coal in its generation portfolio from over 75% to 20%. Note that RE feed-in tariffs were experimentally introduced in 2009, but were quickly phased out and replaced by market-based and competition-favouring programs, namely the REIPPPP competitive bidding framework. As a result, IPPs are beginning to enter the market with renewable projects even though many of them continue to face delays (even those at advances stages) due to financial issues. A regulatory framework to complement the program is expected to be submitted in 2018. In previous years, the introduction of a carbon tax has also been discussed, with the side-objective of rendering renewables deployment more attractive. However, the introduction of the tax has been deferred several times as it would represent a highly impactful policy for the country industrial sector given the large financial implications for ESKOM and the mining sector. It is also worth noting that a biofuels blending mandate was supposed to come into force in 2015, but the government had not released neither the pricing nor the final position paper.

References

Africa Oil Corp. (2016) Africa oil announces significant increase 2C oil resources. http://africaoilcorp.mwnewsroom.com/Files/0c/0c181237-25da-4bb4-84af-bf5bedfe20f1.pdf

CIA (2017) The world factbook 2017. https://www.cia.gov/library/publications/the-world-factbook/

Climatescope - Bloomberg New Energy Finance (2017) Climatescope 2017. http://global-climatescope.org/

ENI (2017a) Volume 2—World gas and renewables review 2017.. https://www.eni.com:443/en_IT/company/fuel-cafe/world-gas-e-renewables-review-2017.page. Accessed 15 Jan 2018

ENI (2017b) Volume 1—World oil review 2017.. https://www.eni.com:443/en_IT/company/fuel-cafe/world-oil-gas-review-eng.page. Accessed 15 Jan 2018

ESCOM (2017) Electricity Supply Corporation of Malawi Limited (ESCOM). http://www.escom.mw/

Gordon E (2018) The politics of renewable energy in East Africa. Oxford Institute for Energy Studies. https://www.oxfordenergy.org/publications/politics-renewable-energy-east-africa/. Accessed 3 Sep 2018

IEA (2017a) World energy outlook 2017

IEA (2017b) WEO 2017 special report: energy access outlook. International Energy Agency

IRENA—REmap (2017) Remap: roadmap for a renewable energy future. International Renewable Energy Agency, Abu Dhabi

Kenyan Energy Regulatory Commission Energy Regulatory Commission. In: Energy Regulatory Commision. https://www.erc.go.ke/

Korkovelos A, Mentis D, Siyal S et al (2018) A geospatial assessment of small-scale hydropower potential in Sub-Saharan Africa. Energies 11:3100

Renewable Energy Atlas Mozambique (2014) Atlas das energias renováveis de Moçambique. http://atlas.funae.co.mz/

RISE (2017) RISE renewable indicators for sustainable energy

The International Journal on Hydropower and Dams (2017) Hydropower and dams in Africa 2017. https://www.hydropower-dams.com/product/africa-map-2017/

The World Bank (2017) World bank data. https://data.worldbank.org/

United Nations (2015) Resolution adopted by the general assembly on 25 September 2015: transforming our world: the 2030 agenda for sustainable development

Chapter 4
Electrification Scenarios

Contents

4.1   Background and Scenarios

In EA-7 (as defined in Chap. 1, i.e. EA excluding South Africa) demand for power is expected to undergo a three-fold increase by 2030, as a result of both electrification (new consumers who gain access), and of increased consumption by already electrified households and by an emerging industrial sector.

According to projections, the compound annual growth rate of electricity consumption is projected at 7.1%, with a total increase from 40 TWh in 2015 to 112.5 TWh in 2030 (Table 4.1). Residential demand will also grow robustly (on average + 10.7% per year). The largest growth rates are expected in countries that currently consume very little power, including Burundi and Rwanda, but in absolute terms the largest increase will be observed in Kenya and Tanzania, the two major economies in the region, where the emerging regional industrial sector will push up the demand.

Mozambique, currently the first consumer in the region due to its power-hungry mining sector (representing 70% of the total demand), displays a lower-than-average expected electricity consumption growth rate (+6.6%). This is owing to the already relatively high per-capita consumption of electrified consumers (264 kWh/capita/year) with respect to the rest of EA-7 (which averages at 187 kWh/capita/year). Further reasons include the weak grid transmission system in place and the large extent of the country, which together are prone to determine a high penetration of decentralised electrification solution and thus a lower per-capita consumption.

Table 4.1 Current and estimates of power demand in EA-7 countries

Country	Gross power demand in 2015 (GWh)	Residential power demand in 2015 (GWh)	Gross power demand in 2030 (GWh)	Residential power demand in 2030 (GWh)	Compound annual growth rate (gross) (%)	Compound annual growth rate (residential) (%)
Burundi	300	150	3500	1500	17.8	16.6
Kenya	9500	2550	23,000	10,000	6.1	9.5
Malawi	2000	600	8000	2500	9.7	10.0
Mozambique	13,500	1650	34,000	4500	6.4	6.9
Rwanda	500	400	4000	2500	14.9	13.0
Tanzania	11,000	2500	30,000	13,000	6.9	11.6
Uganda	3250	743	10,000	4500	7.8	12.8
Overall	40,050	8600	112,500	38,500	7.1	10.5

Sources: Authors' elaboration on (Lahmeyer International and Electrogaz; Mahumane et al. 2012; SEforALL 2013; Zalengera et al. 2014; Ministry of Energy and Mineral Development 2015; Mawejje and Mawejje 2016; CIA 2017; IEA 2017a, b; Keizer 2017; Teske et al. 2017)

Table 4.2 Grid electricity generation (GWh) scenarios in EA-7

	Scenario	Hydro (%)	REs (%)	Natural gas (%)	Coal, diesel, and HFO (%)
2015	Total generation[a]	62	13	12	13
2030	Scenario 1 (Hy + NG)	55	10	25	10
	Scenario 2 (Hy + Co)	50	10	10	30
	Scenario 3 (Hy + RE-NG)	50	22.5	22.5	5

[a]Only electricity consumed in the region is considered. Thus, the ~13 TWh/year of electricity generated at Cahora Bassa dam in Mozambique but exported directly to South Africa are excluded from the calculation

In this context, two questions are deemed very relevant to address, namely:

1. with which generation sources can the growth in the overall demand be satisfied;
2. which is the least-cost way of delivering access to the EA-7 population without access.

Concerning the generation mix, currently (as seen in the first row of Table 4.2):

- The bulk of the installed capacity of EA-7 is given by hydropower, with medium and large-sized dams (overall 3 GW) providing almost the entire power supply of the countries under analysis. Note that at least a further 15 GW of capacity have been planned.
- Other renewables (solar, wind, geothermal, and biomass) display a penetration of 10%, with 600 MW of geothermal capacity in Kenya, and some wind (e.g. the 310 MW Lake Turkana farm in Kenya) and solar (e.g. the 10 MW Tororo station in Uganda) recently, or being in the process of, coming on-line.

- Gas-fired generation is significant only in Tanzania (more than 700 MW operating). Diesel and HFO account for further 500 MW in EA, and they play an important role in Kenya, where they account for roughly 25% of national generation.
- Installed coal generation capacity is currently very limited (0.25 GW of installed capacity). Significant plans exist for developing coal power plants in different countries, including Kenya and Mozambique.

To discuss the potential evolution of the regional grid-based electricity generation mix up to 2030 (the planning horizon adopted in this analysis, as well as that of SDG 7), three scenarios have been designed.

- Scenario 1 represents a trajectory of slightly reduced dependency from the predominance of hydro (with capacity additions of up to +12.5 GW), implementation of on-grid REs (+1.6 GW) and some coal (+1.1 GW), and the bulk of new non-hydro capacity based on NG (+4.6 GW). It is a scenario where EA-7 NG resources are developed for domestic use and a NG pipeline distribution network begins to be developed across the region. Overall, the shares of hydro, coal, and REs diminish, as in relative terms capacity additions are lower than those of NG and of the increase in electricity demand.
- Scenario 2 describes a path where NG is partially devoted to exports out of the region. Its overall share remains constant (with +1.4 GW added), while the bulk of non-hydro capacity additions is coal-based (+4.5 GW), with both coal imports (from South Africa, DR Congo and Zimbabwe), and coal-mining activity (in Tanzania and Mozambique). As in Scenario 1, the share of REs for on-grid decreases slightly over the period (+1.6 GW). Hydropower continues to represent the majority (+50%) of total capacity (+11 GW added).
- Scenario 3 is a pathway of rapid REs (+4.8 GW) uptake in tandem with NG production for domestic generation (+3.8 GW), where coal share remains constant (5%, i.e. +0.25 GW) with respect to the current share, and hydropower share is slightly less prominent (+11 GW), with a final configuration of roughly 50% for hydro and 22.5% each for NG and REs.

The projected levelized costs of electricity (LCOE)[1] generation for each technology up to the year 2030 are reported in Table 4.3. Figures take into consideration expected cost profile changes of the different generation technologies. For NG (the only generation technology considered for which the fuel costs makes a highly significant difference in rendering it competitive or less) we estimated—based on Demierre et al. (2015) and expert assessment—that the domestic gas price of gas in EA until 2030 will be around 4–5 USD/Mbtu, and therefore that the average LCOE of gas-fired generation will stand at 0.06 USD/kWh.

---

[1]"The cost of supplying a unit of energy over a system's lifetime that incorporates the initial investment in generation, transmission and distribution infrastructure; capital costs; and operations and maintenance costs including fuel costs. Levelized costs allow us to compare different technologies on the basis of the minimum unit price a user must pay for each system to break even". (Deichmann et al. 2011).

Table 4.3 Average LCOE
between 2015 and 2030

Technology	LCOE (USD/kWh)
Hydro	0.04
REs (solar PV, geothermal, and wind average)	0.05
NG	0.06
Coal	0.08

Sources: Projections based on data from IRENA (2018) and Santley et al. (2014) and on forecasts comparing estimates from Creutzig et al. (2017), IRENA (2016a, b), Varro and Ha [IEA, NEA, OECD] (2015), and Augustine et al. [NREL] (2018)

Table 4.4 Average cost of
grid-based power generation
(2015–2030) under the
generation mix scenarios

Scenario	Baseline cost of grid electricity generation (USD/kWh)
1	0.05
2	0.055
3	0.049

Starting from these LCOE figures, the average cost of grid electricity generation[2] for each scenario has then been calculated, and they are reported in Table 4.4.

Finally, levelized costs have been adjusted on a country-by-country basis to account for the different endowment of energy resources. This implies that countries which have local abundance of a given energy resource will display lower costs in scenarios where such resource is largely exploited. In particular:

- For Scenarios 1 and 3, which determine a higher penetration of NG-fired generation, no additional costs accrue for Tanzania and Mozambique (which are endowed with reserves), while a cost premium for other countries which would have to import such resources is added.
- In Scenario 3, where RE are also prominent, these are rendered relatively costlier in smaller countries because it is assumed that their size will prevent them to achieve the scale dynamics associated with the greater potential of larger countries.
- Finally, in Scenario 2, where coal-fired generation gains a significant share, smaller countries more distant from coal-bearing areas (Mozambique, Zimbabwe and DR Congo) face a higher price.

The next Sect. (4.2) will be devoted to an analysis of the least-cost ways to bring access to electricity to the populations that currently have no access to electricity in

---

[2]It must be noted the cost of grid-based power generation is the cost borne by the electricity utility for power generation, not the price paid by the end-user. It does not include transmission and distribution network costs, which generally represent a larger fraction of the total cost of electricity delivered to end-consumers, or taxes and subsidies on consumption.

EA-7. Subsequently, Sect.4.3 will estimate (1) the investment required and (2) costs for satisfying the growth in the demand for power from the population that has currently access as well as from the industrial sector.

4.2   Assessing Least-Cost Electrification Options for the Population Without Access

An electrification model (the OnSSET, Open-Source Spatial Electrification Tool, developed by the Department of Energy Systems Analysis at the KTH Royal Institute of Technology; see Mentis et al. 2017) is used to compute the required capacity and investments needed to attain a least-cost 100% electrification by 2030 in all EA-7 countries (as in compliance with the UN's Sustainable Development Goal 7). South Africa is excluded from this analysis due to the already high level of access to electricity in the country.

A total of 12 scenarios that vary across three key dimensions are considered, namely:

1. The baseline price of diesel in 2030, for which a low-price (0.90 USD/l) and a higher-price (1.30 USD/l) are defined.[3]
2. The grid-based power generation mix scenario (as defined in Table 4.2 above).
3. The electricity demand tiers in urban and rural areas, respectively. A high-tier scenario (with 423 and 160 kWh/person/year in urban and rural areas) and a low-tier scenario (with 160 and 44 kWh/person/year in urban and rural areas) are defined.

To provide a sense of what such figures mean:

- 44 kWh/person/year (low-tier in rural areas) are enough to provide general lighting, air circulation and a television;
- 160 kWh/person/year (low-tier in urban areas and high-tier in rural areas) also enable some light appliances use, such as general food processing and washing machine;
- 423 kWh/person/year (higher-tier in urban areas) further include medium or continuous appliances, such as water heating, ironing, water pumping, refrigeration, and microwave.

The high-tier scenario could thus be thought as an evolution from the low-tier scenario, i.e. as a scenario which includes not only the provision of basic household electricity, but also the additional demand for power from the rise of handicraft, the opening of small businesses, and the gradual growth over time in power consumption from newly electrified households.

---

[3]The current average price in the region is above $0.80/l (GIZ 2017).

Note that in the analysis a discount rate of 10% has been considered. The discount rate is an important parameter in determining the results of scenario analysis, since it is the factor which measures the rate at which a society is willing to trade present for future consumption (i.e. costs and benefits in the present and in the future). The decision to set it precisely at 10% was made from Pueyo et al. (2016)'s treatise on the discount rate in RE projects in Sub-Saharan Africa. The authors refer to the fact that the World Bank typically adopts a social discount rate of 10 per cent to assess infrastructure investments in developing countries, and that—for instance—cost-benefit analyses in Kenya use a social discount rate of 10%.

Figure 4.1 reports the average country-level required investment (including investment for both capacity additions and the installation of the necessary transmission and distribution grids) for providing least-cost access to the entire population of EA-7 by 2030. On the left-hand-side, the numbers refer to the low-tier consumption objective, while on the right-hand-side they imply substantially higher per-capita demand of newly electrified households. Overall, the average total required investment for EA-7 stands at $86 billion, or $5.8 billion/year until 2030. Switching from low to high consumption tiers implies—on average—an 85% increase in required investments (from $61 to $113 billion). Results show that Tanzania, Kenya and Uganda are the countries with the largest overall investment required to electrify households without access by 2030. Between today and 2030, the three countries will have to electrify 72 (Tanzania), 53 (Kenya), and 55 (Uganda) million people if they want to attain full electrification (accounting for population growth).

With regards to the optimal technology mix, Figs. 4.2 and 4.3 report for each generation technology the range of required capacity additions and investments in each country across the scenarios considered. In the graphs, grid refers to electrification by grid-connection, MG defines mini-grid systems, and SA refers to decentralised stand-alone solutions.

As shown in Fig. 4.2a, Tanzania and Uganda together require around half (3.5 GW) of the total (7 GW) new grid-connected capacity additions for delivering 100% electrification to those without access in EA-7. Kenya and Mozambique require a median of 1 further GW each. Interestingly, the range of uncertainty across scenarios is relatively little for Kenya and Mozambique (0.6–1.4 GW in both countries), while it becomes larger for Tanzania (1–2.8 GW) and Uganda (0.7–2.4 GW). This implies that in Kenya expanding access through grid electricity remains a relatively efficient solution as opposed to decentralised solutions under all costs and demand scenarios analysed. The result stems from the fact that among EA-7 countries, Kenya has already the highest electricity access rate (56%, with 77% in urban and 39% in rural areas). Thus, relatively little grid-based generation capacity is required to feed the remaining non-electrified centres.

Mozambique, which is the second largest country in terms of total surface area after Tanzania, requires less than half of the median grid-based generation capacity required by Tanzania. This is due to its population being only half of that of Tanzania, to the very low population density of 37 people/km^2 in Mozambique as opposed to a density of 63 people/km^2 in Tanzania, and to the weak electricity transmission network in place. In Mozambique, limited interconnections exist

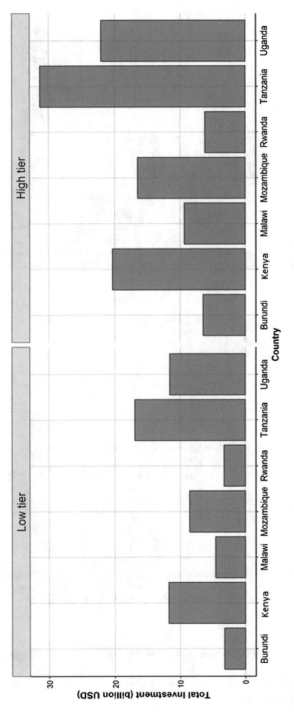

Fig. 4.1 Aggregated required investment by country and consumption tier for newly electrified households

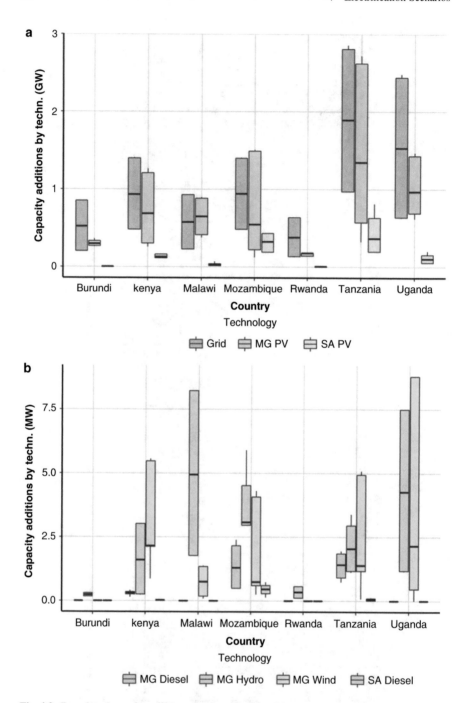

Fig. 4.2 Box plot of capacity additions for electrification of those currently without access by 2030 (by technology and by country)

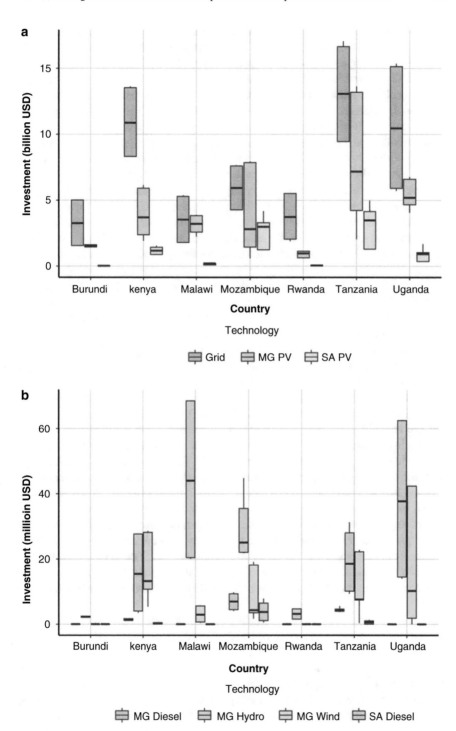

Fig. 4.3 Box plot of total required investment for electrification of those currently without access by 2030 (by technology and by country)

between the south, in proximity of the capital city Maputo, the centre, around the Zambesi river basin, and the north, where currently the bulk of the transmission grid is installed. The massive Cahora Bassa dam, located in central Mozambique, is only connected to transmission lines to South Africa and towards the north of the country. Such conditions render grid expansion only economically feasible in high density provinces, and in turn make standalone (SA) and mini-grid (MG) all very significant generation solutions to fully electrify the country at the lowest cost.

In small-sized but densely populated EA-7 countries, namely Burundi and Rwanda, grid electricity takes up the bulk of access expansion, with 0.5 GW of median on-grid capacity additions in each country (Fig. 4.2a). Also in Malawi around 0.5 GW of on-grid capacity is required, but in addition mini-grid solutions also have a prominent role in realising least-cost electrification (with median required capacities of 0.6 GW for PV and 5 MW for hydro, see Fig. 4.2a, b).

Comparing model results for capacity additions (Fig. 4.2) and investment required (Fig. 4.3), it is interesting to observe how the required investment for grid electricity in Kenya ($12 billion) is similar to that of Tanzania ($14 billion) and Uganda ($11 billion), irrespective of the notably lower capacity addition required (see Fig. 4.2a). This is due to the fact that grid expansion will constitute a major source of costs for achieving 100% access in Kenya, as the figures of required investment include investment in new transmission grid capacity.

Furthermore, important MG solar investments are required in Tanzania ($7 billion) and Uganda ($5 billion), while SA PV investment is prominent in Tanzania ($4 billion) and Mozambique ($3 billion). Concerning minor technologies (in terms of their penetration), MG hydro will have the greatest required investments in Malawi and Uganda (around $40 million in each), countries with substantial potential and conditions for small-scale hydro development, while it will also require $25 million in Mozambique and $19 million in Tanzania. MG wind requires around $ten million in each of Kenya, Tanzania, and Uganda. The only country where SA diesel will have a notable required investment, namely $5 million, is Mozambique, where—in particular in non-electrified but densely-populated coastal areas—it would be the least-cost option due to the low fuel transportation costs.

The regional least-cost electrification situation is outlined in the maps of Fig. 4.4. These provide a visual insight into the technology with the lowest levelized cost of electricity in each square-km area of EA-7 given the electrification tier set under the scenario parameters in examination. Two cases are reported, namely the high and low-consumption tier variants of Scenario 3 (high penetration of renewables and NG).

It can be observed that independently from the case considered, in the vast majority of the region solar mini-grid and off-grid are the least-cost options, while in coastal areas diesel is a viable option due to the very low transportation costs it faces there. Grid electricity is least-cost in large urban areas, mostly in the northern part of EA. Wind and hydro mini-grids are instead only cost-competitive in very circumscribed areas of high potential and population density. A higher electrification tier increases the number of locations where on-grid electrification is least-cost and it

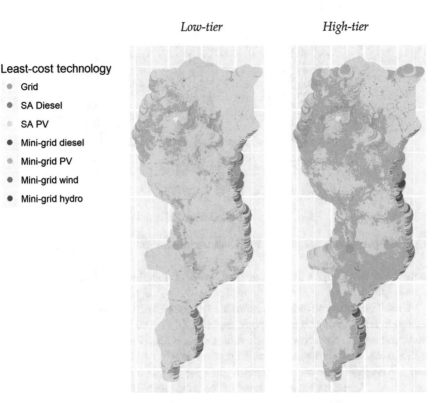

Fig. 4.4 Maps of least-cost technology across EA-7 for Scenario 3

boosts the relevance of solar MG vis-à-vis that of solar SA. Maps for all the remaining scenarios are found in *Appendix B*.

As can be seen in Fig. 4.5, to achieve full-electrification, grid generation capacity additions and its extension are always the largest investment component, with a mean value of 58% of total required investments, while MG mean investments stand at 34% and SA mean investments at 9%. Total corresponding capacity additions across the three scenarios stand at 13 GW, roughly split into 50% of grid capacity, 41% of mini-grids (including PV, wind and hydro-based solutions) and 8.5% of standalone solutions.

Results hence show that mini-grid solutions will have great importance in fostering electrification, irrespective of variations in the price of grid electricity. This is owing to two key facts, namely (1) that in most EA-7 countries the national grid is still weakly developed and the cost to be borne to extend it are massive and not efficient, at least if the objective is that of achieving the tiers of electrification which were considered in our modelling exercise; (2) that those without access are often remote for the national grid but concentrated in medium/high density settlements where a joint generation system is more cost-effective than each individual

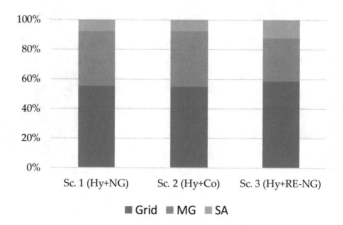

Fig. 4.5 Split of grid, mini-grid, and standalone overall required investments, by scenario

household installing a private device, even at low levels of per-capita power consumption.

It must be remarked that such estimates are conservative, in the sense that they are served to reach at most a tier of consumption of 423–160 kWh/person/year in urban and rural areas, respectively. Such values remain low if compared with the current average residential per-capita consumption levels of South Africa, standing at 807 kWh/capita. The model output figures are in fact addressing the challenge of bringing a minimum guaranteed level of access to electricity to the entire population. More realistic estimates, tailored to the specific needs of each region within each country, should be addressed by country-level or sub-national studies.

The average investment required by 2030 to provide universal access to electricity in EA-7 stands between $113 and $61 billion for high and low-tier consumption levels, respectively. This averages at $87 billion USD, which corresponds to roughly 5.8 billion/year in the model's planning horizon between 2015 and 2030. If we take as a benchmark the mean full electrification cost of USD 21 billion per year estimated in the PBL Netherlands Environmental Assessment Agency's *Towards universal electricity access in Sub-Saharan Africa* report (2017) and we divide it by the share of the EA-7 population over the total SSA inhabitants (i.e. 21%), we find a value of 4.4 billion USD/year. This suggests that the two figures are relatively similar, and that analogous results are achieved using different models, parameters, and methodology.

Finally, along with a comment of the results, it is meaningful to describe the main limitations of the least-cost electrification exercise, so that readers can interpret the results in a meaningful way. The different scenario results are informative in terms of the comparative information they provide, rather than the absolute output values they suggest. Comparing scenarios with different parameters can shed light on the relative significance of different factors in determining the optimal power mix to achieve least-cost electrification. The model itself embeds a long-list of assumptions and of a priori set parameters which might not necessarily reflect the actual state of things in the different EA-7 countries under examination, even if we have accounted for some of the specific characteristics of each country in setting the relative variables in each scenario run. The same holds for projections over the evolution of demographic and cost variables up to 2030. Another caveat stems from the fact that the model does not account for non-electric needs (e.g. clean cooking), but it simply estimates the least-cost way of providing universal access to electricity within a country. For governmental planning purposes, a more in-depth and detailed study, based on field-data assessment, would need to be carried out.

4.3   Beyond Access: Scenarios for Satisfying the Demand Growth of Already Electrified and Industrial Consumers

In EA-7, additional demand growth will represent a further significant challenge beyond the provision of universal basic access to electricity. Demand for power is projected to grow at an average yearly rate of 7.1% as a result of both a greater per-capita consumption from those who already benefit from access today and of increased industrialisation and mechanisation of agriculture.

Therefore, we calculate the up-front investment required to match the growth of the demand for power generation of those already electrified (including from non-residential sectors) over the three grid electricity generation mixes scenarios introduced in Table 4.2. Investment costs for each technology in Africa are drawn from Enerdata's *Study of the Cost of Electricity Projects in Africa* report (2016).

The mean required investments range is within a rather narrow range (Fig. 4.6), i.e. between $41.2 billion for a scenario of reduced diversification from hydropower (Scenario 1) and $38.7 billion for a RE-NG expansion scenario (Scenario 3), with the coal-based expansion scenario (Scenario 2) requiring $39.1 billion. It is then

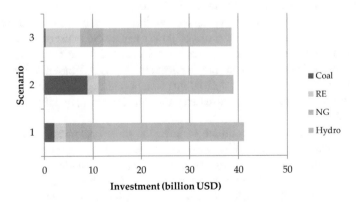

Fig. 4.6 Required up-front investment in capacity for non-electrification demand growth

Table 4.5 Cumulated grid electricity generation cost from already electrified demand growth (2015–2030)

Country	Scenario 1, demand growth only (bn. USD)	Scenario 2, demand growth only (bn. USD)	Scenario 3, demand growth only (bn. USD)	Mean across scenarios, demand growth only (bn. USD)
Burundi	0.14	0.15	0.13	0.14
Kenya	3.44	3.45	3.22	3.37
Malawi	0.34	0.34	0.33	0.34
Mozambique	0.61	0.67	0.59	0.62
Rwanda	0.15	0.16	0.15	0.15
Tanzania	1.02	1.23	1.00	1.08
Uganda	0.52	0.58	0.51	0.54
Total	6.20	6.58	5.93	6.24

clear that in investment terms there is very little difference across the three scenarios considered.

Overall, in all scenarios the bulk of the required investment to cover the additional demand in EA-7 (i.e. not stemming from new connections) is for hydropower capacity. At the same time, the projections show that the investment required for capacity addition in order to cover the demand growth from connected customers are on average around 50% of those needed for expanding electricity access (which however also include investment to expand the national grid where that is necessary to deliver electricity).

Table 4.5 reports instead the total power generation cost which will need to be covered to satisfy the baseline consumption growth (i.e. not stemming from newly electrified consumers) until 2030 under each scenario. Generation costs, differently from investment, are not the component required up-front, but rather the overall cost required over time, taking into consideration the investment component, the fuel component and the operation and maintenance components (i.e. the LCOE multiplied by the amount of electricity generated between 2015 and 2030).

Results show that for all countries Scenario 3—that of a capacity expansion based on RE and NG—is the cheapest solution to satisfy grid-based electricity demand. Scenarios 1 and 2, i.e. those of power expansion backed by NG or coal, respectively, show instead higher costs, with Scenario 2 being the costliest (+10% vs. Scenario 3). According to our results, Scenario 3 (RE-NG-based expansion) will be both the one with the lowest upfront investment (although this will be very close to the upfront investment of a coal-based expansion scenario), and the cheapest over the long-run.

While hydropower will continue to have the most prominent role, the future power mix will depend on political choices taken over the next decades. Overall, according to the scenarios elaborated in this analysis Scenario 3 (RE-NG-based expansion) will be both the one with the lowest upfront investment (although this will be very close to the upfront investment of a coal-based expansion scenario), and the cheapest over the long-run. RE-NG development (Scenario 3) will guarantee

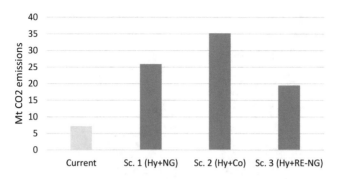

Fig. 4.7 Carbon dioxide (CO_2) emissions in 2030 by generation mix scenario

lower future costs (-10% than Scenario 2 and -4.5% than Scenario 1), while it would also contribute to attain a lower (greenhouse gas emissions pathway (-45% vs. the coal-based expansion of Scenario 2 and -25% vs. the gas-based expansion of Scenario 1 in year 2030, as seen in Fig. 4.7) and local pollution levels.

Carbon dioxide emissions are the first driver of human-induced global warming and climate change, for which multiple studies (see IPCC 2014) have predicted an adverse socio-economic impact on developing economies of Sub-Saharan Africa, where the financial capacity for adaptation is limited. The projected emissions of EA-7 by 2030 under all scenarios are negligible if compared to the current global emissions (they would represent a share of around 0.11% of total current emissions). Nonetheless, a more sustainable development path—like that of Scenario 3—still allows to reduce the social costs of power generation (even locally, for instance with the emission of less local pollutants such as those resulting from coal combustion). Note that however such external costs have not been included in this analysis.

4.4   Investment Requirements in Perspective

The results of our analysis in the previous sections show that the investment so that generation capacity addition matches demand growth beyond electrification (standing at a mean of \$40 billion) are around half of those needed for new electrification itself (namely \$87 billion for a mean level between low and high-tier consumption). However, the figure of investments for newly electrified consumers already includes grid extension investment where this is necessary to deliver electricity, while that for already electrified consumers only includes power plants additions.

Assuming that—in first approximation—transmission and distribution investments needed to accompany the power generation expansion are generally similar to those of the generation investment itself (IEA 2016), we derive for the grid based investments a total investment cost for power generation, transmission, and distribution of about \$40 billion dollars between 2016 and 2030. Summing them to

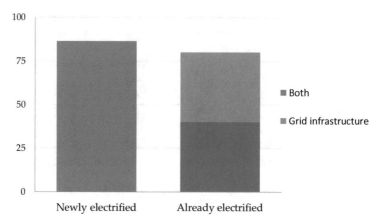

Fig. 4.8 Investment requirements for newly and already electrified consumers in EA-7

investments for new electrification, a total requirement of $167 billion in a time span of 15 years (2016–2030) will be needed for EA-7. Figure 4.8 highlights this key finding.

Thus, it results that the two typologies of investment are similar in magnitude. Yet, the two cannot be directly compared, since they present some substantial differences in terms of the underlying financing dynamics. While investments for already electrified and non-residential consumers will mostly be sustained directly by the consumers of such additional power (via bills), the electrification investment for new consumers will be instead affected by issues on inability-to-pay for the upfront investment required, in particular for grid-connection charges. As a result, the two investment requirements will need to resort to different sources and modes of financing, which are discussed later in this book (see Sects. 5.4 and 5.5).

To put results in perspective, the required investment per-capita and as a share of GDP (both current and PPP) have been calculated for each country and in each year. In doing so, it was assumed that investment will not be evenly distributed across the 15 years under consideration, but rather that it would increase in an exponential fashion, namely at the same rate at which electricity consumption is projected to grow between 2016 and 2030 in each country. Figure 4.9 plots a representation of such evolution of total required investment over time for both newly electrified and already electrified consumers in EA-7.

Country-level investment figures for newly electrified consumers are reported in Table 4.6. They refer to the mean value between low and high-tier levels of consumption. The table also shows the required investment as a share of GDP (in both the exchange-rate, PPP, and weighted formulations) in 2016, and average per-capita per year within each country.

Results suggest that the average per-capita investment (accounting for the entire population, even those who already have electricity access) for electrifying households currently without access stands in the range of $19–25 per capita, with the regional EA-7 figure at $22. When calculating the same figure but only for the

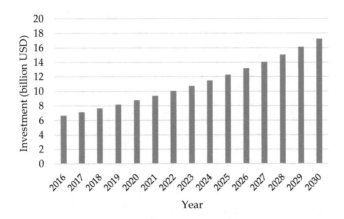

Fig. 4.9 Representation of the exponential growth of total required investments between 2016 and 2030

population currently without access (thus what hypothetically each person should finance himself to gain access), the figure rises to a range of $33–42 per capita, with however only marginally higher values in many countries (as a result of low-electrification rates, which make the total population and the population without access similar). This represents roughly 1.9% of the national GDP in exchange rate terms in the initial year (i.e. 2016), or 0.7% in PPP terms.

Whether to consider the exchange rate or the PPP GDP as the reference metric depends on the destination of the electrification investment, i.e. whether the installed infrastructure and the corresponding labour will be of domestic or foreign origin. According to experts' assessment, in EA-7 roughly 25% of the investment is expected to target the local industry, labour and public sector, while the remaining 75% will finance international acquisitions of technology and materials and international corporations operating in EA-7 countries. In particular, hydropower, thermal, PV, and transmission and distribution grid will be largely constituted by foreign assets, especially hard infrastructure built abroad. Thus, it might be reasonable to consider a weighted formulation of GDP, which is given by the weighted average of PPP GDP for international-targeted investment and of exchange rate GDP for local-targeted investment. Such metric is reported in the last column of Table 4.6 (for newly electrified consumers) and of Table 4.7 (for already electrified consumers).

Again, it must be remarked that for newly electrified consumers we are considering a mean value between high and low tiers of consumption as defined in Sect. 4.1. If we look at the actual results for each of the two tiers, the investment requirements stand at 1.3% (low-tier) and 2.4% (high-tier) for the exchange rate GDP, at 0.5–1% of the PPP GDP, and at 0.9–1.6% of the weighted GDP. Thus, increasing new electrification efforts from the low to the high tier would result in around a doubling of investment requirements for bringing access to households currently without power. As discussed in Sect. 4.1, the high-tier scenario can be thought as an evolution from the low-tier scenario, i.e. as a scenario which includes

Table 4.6 Investment requirements for newly electrified consumers in perspective

Country	Total investment (bn. USD)	Average investment per capita per year (USD)	Average investment per inhab. without access per year (USD)	Investment as a share of 2016 GDP (%)	Investment as a share of 2016 PPP GDP (%)	Investment as a share of weighted[a] GDP in 2016 (%)
Burundi	4.85	23	37	2.3	1.0	1.7
Kenya	16.1	19	33	0.9	0.4	0.7
Malawi	7	21	35	3.6	1.1	2.3
Mozambique	13	25	42	4.4	1.5	3.0
Rwanda	4.70	23	40	0.2	0.4	0.2
Tanzania	24	23	39	3.7	0.6	1.7
Uganda	16.8	22	36	6.9	0.8	2.4
Total	87	22	37	1.9	0.7	1.3

[a]We define weighted GDP as the weighted average of PPP GDP by the share of international-targeted energy access investment and of exchange rate GDP by the share of local-targeted investment

Table 4.7 Summary of investment requirements for each consumers category

Category	Investment (bn. USD)	Average investment per capita per year (USD)	Investment as a share of 2016 GDP in 2016 (%)	Investment as a share of 2016 PPP GDP in 2016 (%)	Investment as a share of weighted GDP in 2016 (%)
Newly electrified consumers	87	22	1.9	0.7	1.4
Already electrified consumers	80	20.5	1.7	0.7	1.2
Total investment	167	42.5	3.6	1.4	2.6

not only the provision of basic household electricity, but also the additional demand for power from the rise of handicraft, the opening of small businesses, and the gradual growth over time in power consumption from newly electrified households. All these activities would contribute to rural development, and thus to increasing the national GDP. Further macroeconomic analysis—in the form of benefit-cost analysis (BCA)—would be required to assess the economic significance of performing the larger investment to provide high-tier consumption access.

Table 4.7 shows—in the first row—results on a regional level for the investment required to satisfy the additional demand growth of already electrified consumers, and—in the second row—investment for total electrification (given by the sum of required investment from newly electrified and already electrified consumers). While results for already electrified consumers are very similar to those for new electrification, total investments results for EA-7 as a whole stand at an average of $42.5/capita per year, which correspond to 3.6% of EA-7's exchange rate GDP, 1.4% of PPP GDP, and 2.6% of the weighted GDP in year 2016.

To put such figures into perspective with the regional socio-economic and financial context, a number of national statistics for EA-7 countries is reported in Table 4.8, including the savings rate, the poverty rate, per-capita income, and information on the Official Development Assistance (ODA) and on the net inflows of Foreign Direct Investment (FDI), as well as on the current yearly investments in the power sector with private participation (where available). For such indicators, which by their own very nature tend to be volatile (e.g. sometimes due to a single mega-project), we considered the average between 2010 and 2017. Such figures enable a better understanding of the current situation vs. the projected trends in EA-7 countries, and thus of the challenges and opportunities to satisfy the investment requirements for electrification.

We observe that in general EA-7 countries display relatively high savings rate as a share of GDP (with a high of 28% in Mozambique, and significant shares in Tanzania and Uganda). Only Burundi displays negative savings. While in an ideal economy savings is equalised to investment, in the context of EA-7 a large part of savings is detained by individuals and companies as a result of the unsuitable and

Table 4.8 Country-level relevant economic indicators

Country	Gross national saving rate (% GDP) in 2017	Population below poverty line in 2017 (<$700/person/year) (%)	GNI[a]/capita in 2017	Net ODA ($) received per capita (USD), 2010–2017 average	Net ODA as a share of GDP, 2010–2017 average (%)	FDI (net inflows ($) per capita), 2010–2017 average	FDI (net inflows as a share of GDP), 2010–2017 average (%)	Investments in power energy with private sector participation (bil. USD)[b], 2010–2017 average	Investments in power energy with private sector participation (share of FDI) (%)
Burundi	−5.7	65	770	58	21	3.7	1	–	–
Kenya	10.6	36	3250	55	4	18.2	1	0.46	55.5
Malawi	3.7	51	1180	63	17	22.8	6	–	–
Mozambique	28	46	1200	75	15	146.8	29	0.22	5.6
Rwanda	12.5	39	1990	97	14	22.1	3	0.11	44
Tanzania	24.1	23	2920	54	7	32.1	4	0.13	7.9
Uganda	20.4	21	1820	44	7	22.4	4	0.04	4.7
EA-7 mean	17.4	34	2250	58.5	7.7	39.8	4.9	0.19	–

Source: The World Bank (2017)

[a]Gross National Income, defined as gross domestic product (GDP), plus factor incomes earned by foreign residents, minus income earned in the domestic economy by non-residents

[b]"Investment in energy projects with private participation refers to commitments to infrastructure projects in energy (electricity and natural gas: generation, transmission and distribution) that have reached financial closure and directly or indirectly serve the public" (The World Bank 2017)

insecure investment environment, therefore slowing economic growth and development prospects.

For example, a savings rate of 28% of GDP in Mozambique denotes that in 2016 around $3.5 billion have been saved, a figure 6×-times larger than the required electrification investment for households without access in that year. At the same time, the very high share of the population living below the poverty line in Mozambique (46%) suggests that the bulk of such savings is owned by a small share of the population in the country, a finding which is underlined by Mozambique's Gini index of income inequality,[4] standing at 0.46 (well-above the world's average). Thus, observing the GNI per-capita figure is representative in so-far it gives a sense of the different income level across EA-7 country, and yet it reveals little about the ability-to-pay of households to afford connection to the national grid or the purchase of standalone solutions.

Other interesting insights come from the observation of Official Development Assistance (ODA), both in per-capita and as a share of GDP terms. ODA is particularly relevant in small economies such as Burundi and Malawi, while more developed economies such as Tanzania and Kenya only received 7% and 4% of their GDP, respectively, or around $55/capita. A different situation is instead outlined when looking at the flows of Foreign Direct Investment (FDI): here, the largest inflows have been towards Mozambique (FDI is as large as 29% of the national GDP), in all likelihood driven by the energy resource-abundance found in the country (with both considerable coal and NG reserves under exploitation). Nowhere else in EA-7 such high rates are found. The second-highest FDI as a share of GDP is in Malawi, followed by Tanzania and Uganda, with the other countries exhibiting figures between 1% and 3%. However, the per-capita terms FDI figures are more significant, as they are not biased by small-sized economies. While Mozambique remains by far the first destination of foreign investment ($147/per capita), Tanzania ($32/per capita) emerges as the second destinations over the last decade. Interestingly, the average figure (excluding Mozambique, since it represents an outlier) stands at $20, and it is not far from the per-capita required total electrification investment.

Nonetheless, the FDI figures alone do not tell much about the destination and the end-use of such investments. In some instances, such as for NG in Mozambique until today, investments may be almost totally serving export purposes, and living little wealth or room for development prospects in the country. For five countries, the World Bank reports the absolute figure in USD for the investments in power energy with private sector participation,[5] from which we also derive the number as a share of FDI. This is an interesting metric to understand the liveliness of the investment sector

---

[4]The Gini coefficient measures the inequality among values of a frequency distribution (in this case family income), where a value of 0 expresses perfect equality and a coefficient of 1 expresses maximal inequality.

[5]Defined as *"commitments to infrastructure projects in energy (electricity and natural gas: generation, transmission and distribution)"*, thus excluding upstream resources production.

and its significance on the overall flows of foreign investments. However, it must be born in mind that—as referred by Eberhard et al. (2017)—over the last 25 years half of the total investment in power generation plants in Sub-Saharan Africa has stemmed from governments and utilities, while the reminder has been split between 22% of IPPs, 16% from China, and 11% from ODA and FDI. This—combined with subsidies on electricity rates—has resulted in large governmental and utilities deficit, which call for larger private sector participation in the investment effort.

In particular, in Mozambique the average investments in the power sector with private participation between 2010 and 2017 amounted to only $0.22 billion, i.e. 5.6% of the average inflows of FDI, and still 65% below the required investment in 2016 to reach universal access by 2030. In Kenya, private-participated investments in power averaged at $0.46 billion (or 55.5% of FDI), 28% less than the required investment for attaining full-electrification. In Rwanda average investment stood at $0.11 billion, but in this case, they represented around 23% more than the required investment for newly electrified consumers, setting the country on a successful pathway to attain full-electrification even earlier than 2030 and without heavy burden on the public finances. In Tanzania, private-participated investments in power stood at 8% of FDI, i.e. 75% below the level for attaining electrification, rendering the country one of the most public finances-dependent in EA-7. In 2018 the country secured a $0.45 billion loan for power projects from the World Bank. Finally, in Uganda the mean investment stood at $0.04 billion, less than a 5% of the required budget for attaining 100% electricity access by 2030 in the country, thus requiring considerable efforts to scale-up private investment.

Overall, the results of this analysis show that achieving electrification targets set for the year 2030 present substantial costs (a total of about $167 billion for newly electrified, already electrified consumers, and other sectors). If this figure is put in perspective and observed in per-capita or as a share of GDP terms (overall $42.5 per capita or 2.6% of weighted GDP), it seems that investment requirements are more affordable than one could think when looking at the absolute number.

The main roadblock is given by the large share of the population of all EA-7 countries that lives below the poverty line, which is largely the same population currently living without access to electricity. It is in this context that the most significant support not only from national governments, but also from international finance institutions needs to be channelled. Although significant efforts are already taking place, more needs to be done if the ambitious objectives are to be reached without putting a tremendous burden on government finances and thus restraining growth in other sectors.

Satisfying the growth in demand of the already electrified consumers and of an emerging industrial sector seems feasible without large international support. Here, the real challenge is that of attaining a suitable investment environment to IPPs to operate in a competitive market and rapidly expand the national installed capacity, while public utilities can focus on grid infrastructure planning and expansion.

Section 5.4 will discuss which policies will be needed to unlock such investment potential, both domestically and from abroad. Subsequently, Sect. 5.5 will discuss more in detail what role international public financing organisations can play in the process.

References

Augustine C, Beiter P, Cole W et al (2018) 2018 Annual technology baseline ATB cost and performance data for electricity generation technologies-interim data without geothermal updates. National Renewable Energy Laboratory-Data (NREL-DATA), Golden, CO

CIA (2017) The world factbook 2017

Creutzig F, Nemet G, Luderer G et al (2017) The underestimated potential of solar energy to mitigate climate change. Nat Energy 2:17140. https://doi.org/10.1038/nenergy.2017.140

Deichmann U, Meisner C, Murray S, Wheeler D (2011) The economics of renewable energy expansion in rural Sub-Saharan Africa. Energy Policy 39:215–227

Demierre J, Bazilian M, Carbajal J et al (2015) Potential for regional use of East Africa's natural gas. Appl Energy 143:414–436. https://doi.org/10.1016/j.apenergy.2015.01.012

Eberhard A, Gratwick K, Morella E, Antmann P (2017) Independent power projects in Sub-Saharan Africa: investment trends and policy lessons. Energy Policy 108:390–424

Enerdata (2016) Study of the cost of electricity projects in Africa for the African Development Bank

GIZ (2017) Non-alternative facts on international fuel prices in 2016

IEA (2016) World energy investment 2016. https://webstore.iea.org/world-energy-investment-2016. Accessed 19 Oct 2018

IEA (2017a) World energy outlook 2017

IEA (2017b) WEO 2017 special report: energy access outlook. International Energy Agency

IPCC (2014) Climate change 2014: impacts, adaptation, and vulnerability. Cambridge University Press, Cambridge

IRENA (2016a) The power to change: solar and wind cost reduction potential to 2025 /publications/2016/Jun/The-Power-to-Change-Solar-and-Wind-Cost-Reduction-Potential-to-2025. /publications/2016/Jun/The-Power-to-Change-Solar-and-Wind-Cost-Reduction-Potential-to-2025.. Accessed 3 Aug 2018

IRENA (2016b) Solar PV in Africa: costs and markets. http://www.irena.org/publications/2016/Sep/Solar-PV-in-Africa-Costs-and-Markets. Accessed 23 Apr 2018

IRENA (2018) Renewable power generation costs in 2017 /publications/2018/Jan/Renewable-power-generation-costs-in-2017. /publications/2018/Jan/Renewable-power-generation-costs-in-2017.. Accessed 3 Aug 2018

Keizer D (2017) Renewable electricity in Kenya. Master's thesis

Lahmeyer International, Electrogaz analysis and projection of Rwanda's electricity. Demand, final report

Mahumane G, Mulder P, Nadaud D (2012) Energy outlook for Mozambique 2012–2030 LEAP-based scenarios for energy demand and power generation. Acumulacao e Transformacao em Contexto de Crise Internacional, Mocambique

Mawejje J, Mawejje DN (2016) Electricity consumption and sectoral output in Uganda: an empirical investigation. Econ Struct 5:21. https://doi.org/10.1186/s40008-016-0053-8

Mentis D, Howells M, Rogner H et al (2017) Lighting the world: the first application of an open source, spatial electrification tool (OnSSET) on Sub-Saharan Africa. Environ Res Lett 12:085003

Ministry of Energy and Mineral Development (2015) Uganda's sustainable energy for all (Se4all) initiative action agenda

Pueyo A, Bawakyillenuo S, Osiolo H (2016) Cost and returns of renewable energy in Sub-Saharan Africa: a comparison of Kenya and Ghana. IDS

Santley D, Schlotterer R, Eberhard A (2014) Harnessing African natural gas: a new opportunity for Africa's energy agenda?

SEforALL (2013) Burundi: rapid assessment gap analysis

Teske S, Morris T, Nagrath K (2017) 100% renewable energy for Tanzania—access to renewable
    and affordable energy for all within one generation (full report)
The World Bank (2017) World Bank Data.. Accessed 20 Nov 2017
Varro L, Ha J (2015) Projected costs of generating electricity–2015 Edition. France, Paris
Zalengera C, Blanchard RE, Eames PC et al (2014) Overview of the Malawi energy situation and A
    PESTLE analysis for sustainable development of renewable energy. Renew Sust Energ Rev
    38:335–347. https://doi.org/10.1016/j.rser.2014.05.050

Chapter 5
Conditions for RE Deployment and Energy Development

Contents

Irrespective of technical abundancy, RE potential *per se* does not imply a structural and inclusive expansion of energy access and an overall sustainable energy development of EA. Proper technological, economic, institutional, and policy considerations must be made to assess which are the best ways and most apt policies to sustain the exploitation of such potential in the regional context in relation to other energy sources, as well as which roadblocks and challenges are faced. A first meaningful consideration in this sense is that EA is characterised by a strong rural-urban imbalance: the majority of the population lives in poorly interconnected rural communities away from the electricity grid, which serves predominantly densely populated urban centres. While plans to tackle the imbalance are in place in virtually every country (both Kenya and South Africa have achieved notable results in this sense), the issue is not going to be structurally overcome rapidly. Thus, as highlighted by the least-cost electrification scenarios in Chap. 4, when discussing the case for renewables to increase and improve access, a distinction must be made

Table 5.1 Main intervention dimensions and key policy challenges

Technological issues	Economic considerations	Transboundary cooperation	Policy challenges	Investment channelling
RE displacement and path dependency of energy mix and infrastructure	On-grid/ decentralised solutions trade-off	Power generation infrastructure	Public and private investment and the role of IPPs	International public finance institutions
Off-grid technology and storage	Household ability/willingness-to-pay for electrification	The Eastern African Power Pool (transmission)	Subsidies, FiTs, and policy instruments	ODA
Hydropower dependency and climate impact	Uncertainty over future costs and developments	Energy resources sharing and water basins management	Payment schemes and the role of digital technologies	China

between national grid expansion to reach additional shares of the population, and specific decentralised solutions.

In turn, considerations on which solutions are the most cost-effective in each region, which level of power supply is taken as short-term target, which lifetime decentralised options can expect (e.g. before the grid reaches that area), which costs profiles they present (including post-installation and maintenance costs), and which are the most apt policies in such different contexts must be made. RE potential deployment also presents financial and governance issues both on the supply and on the demand-side, with the channelling of private investments being a necessary condition along with the establishment of effective business-to-consumer (B2C) models to ensure consumers are willing and able to pay for energy services.

Moreover, irrespective of large technical potential, renewables alone cannot fulfil all energy needs due (1) intermittency issues and generation fluctuations and (2) non-electric needs by the residential, industrial, and transport sectors. Concerning the first question, round-the-clock power availability requires renewables being complemented by other sources (e.g. through thermal generation at demand peaks) and sustained by storage technologies (for instance with batteries or pumped-storage hydropower) within a sound policy framework. On the second, questions relating to clean cooking and the substitution of solid biomass with liquefied petroleum gas (LPG) and to the exploitation of NG resources to empower the industry and transports in the region are deemed significant.

In this context, five key intervention dimensions are discussed: technological issues, economic considerations, transboundary cooperation, policy, and financing challenges. The main questions touched upon within each cluster are reported in Table 5.1. The discussion is tailored to the specific situation characterising the context of EA, for which relevant experiences and policy cases are reported. Chapter 6 then discusses the role of NG both alone and in relation with RE, and the potential for LPG penetration.

5.1 Technological Issues

5.1.1 RE Displacement and Path Dependency of Energy Mix and Infrastructure

As observed by Fouquet (2016), *"energy systems are subject to strong and long-lived path dependency, owing to technological, infrastructural, institutional and behavioural lock-ins"*. Particularly during the industrialisation and development phases of a country, where flexibility is large and investment in hard assets is heavy, policymakers face the responsibility of carefully considering all implications before directing their economies onto certain energy pathways. For instance, rapid development pathways with a high energy and carbon intensity are prone to be detrimental to their long-run prosperity despite tackling short-run issues. In particular, economies of scale, as well as learning and network effects (determining decreasing marginal cost with increasing installed capacity of a given technology) pave the way towards energy mixes which are not necessarily socially optimal over the long-run. Furthermore, there is a feedback mechanism between energy resources, infrastructure, and industrial development, locking an economy into specific consumption patterns.

The case of coal in South Africa is emblematic in this sense. The country is one of the top world's producers and exporters of coals, which is also used in 90% of its domestic power generation (RISE 2017), with almost 38 GW of installed coal-fired capacity with just around 35% of efficiency (Sloss 2017). Many South Africa's power stations are in the surroundings of a coal mine, from which these are directly supplied with fuel. Coal is also liquefied to satisfy around a third of the domestic demand for liquid fuels by Sasol company (Höök and Aleklett 2010). On the one hand, this setting has been the key driver to the energy independency of South Africa and the rapid expansion of electricity access thanks to cheap domestically mined coal, although recently prices soared. On the other hand, it has rendered the country the 16th world emitter of carbon dioxide, with 0.4 Mt in 2015 (Netherlands Environmental Assessment Energy data) representing alone 1.16% of the global fossil fuel CO_2 emissions. This is striking, especially if compared with the fraction of South Africa's GDP over the global figure, which stands at 0.43%, determining one of the highest carbon intensities of GDP in the world. Also, local pollutants emissions must be factored in, with fossil-fired generation in the country being responsible of large shares of the total SO_2 and NO_x emissions (up to 75% in the Highveld region, one of the key mining areas in the country), resulting in major social costs in terms of air pollution and health impact. Moreover, proven reserves have been significantly decreasing throughout the last 40 years (Fig. 5.1), and as of 2017 they stand at 10 Gt, determining growing concerns for both the domestic energy mix (which has started a massive process of diversification with the displacement of some old-generation coal-fired plants) and the economic growth prospects.

While coal is and will continue to be a very significant energy source for the country in the coming years, it is clear that having put in place an energy system that

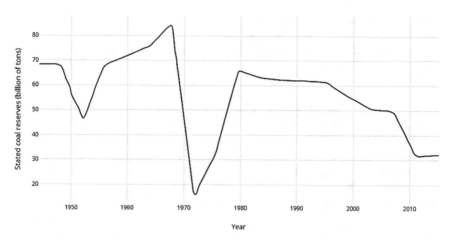

Fig. 5.1 Proven coal reserves in South Africa. Source: Authors' elaboration on World Energy Council (2017)

is almost entirely dependent on it is nowadays raising environmental and economic issues. Thus, virtually all EA countries (which are either not particularly richly endowed with coal or that find themselves in a critical juncture for their energy development pathway) might undergo significant risks if they invest heavily in coal-fired generation capacity. This is true even if in many instances coal-fired generation currently is the one with the lowest short-run LCOE. Significant economic repercussions could be witnessed by such countries as a result of changing prices (Fig. 5.2)—especially if coal is imported—and environmental and health impact in the following decades. The introduction of a global carbon tax—raising the global price of embedded CO_2 and thus of imported coal—is a further risk borne by countries setting up a coal-fired energy development plan.

Upon these considerations, policy makers should carefully consider electrification plans and the means selected to achieve results in their BCA, paying attention not to discount the future disproportionately, i.e. disregarding large future costs despite present benefits. The large-scale set-up of fossil-fired plants to provide baseload power is a relevant example, with future costs accruing from price uncertainty, import dependency issues, as well as climate and health impact. On the other hand, modern RE should be considered by policymakers even though its LCOE (levelized cost of electricity) is higher than other alternatives when the project is drafted, or despite higher upfront deployment costs. RE sources present in fact long-term benefits and steep learning curves (Creutzig et al. 2017; IRENA 2018), with the achievement of energy independency being a major long-term benefit. A challenge for the realisation of RE potential is thus the proper planning of energy development, and thus the avoidance of short-lived investments which would become economically inefficient during their planned lifetime.

Overall, infrastructure projects must be carefully designed and their effectiveness in the coming years under different scenarios (i.e. their resilience) must be assessed:

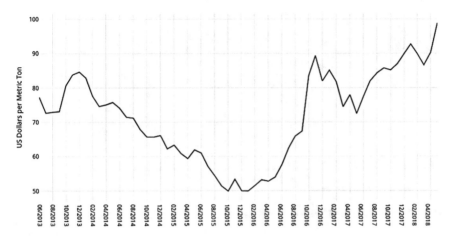

Fig. 5.2 5-year evolution of the price for thermal coal in South Africa. Source: Authors' elaboration on IndexMundi (2018)

if wrongly timed, unevenly spread, or poorly managed, they would eventually foster inefficient outcomes, distributional issues, and long-lasting public debt. Thus, government should carefully consider future investment, weighting pros and cons of large grid-connected projects against those of small and localised ones. At the same time, they must consider how their decision will affect the future path-dependency of the energy system; the potential risk from resources associated with a particular generation technology[1]; scale and network dynamics; expectations about changing costs profiles and technological advances; as well as proper ways to complement the current limitations of generating electricity from modern renewables.

5.1.2 Off-Grid Technologies and Storage

Decentralised solutions demand specific technical considerations. First, policymakers should make projections of where, if, and when the national grid will reach currently non-electrified areas and thus support different RE-backed plans in different areas. Promoting off-grid household-level options or a village-level mini-grid may be inefficient if the grid is to be extended in a short time span, thus resulting in generation capacity redundancy, unless a proper plan to connect decentralised systems to the grid itself are set up. Conversely, where economic and landscape constraints are deemed too heavy, decentralised solutions should be promptly fostered.

---

[1]For example, the availability of water for the cooling of thermal power plants or possible hydropower disruptions in areas prone to suffer from drought events.

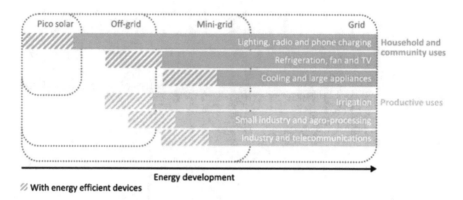

Fig. 5.3 Means of electrification and their possible uses. Source: IEA (2017a)

Second, the tier of electricity (expressed e.g. by the kWh/year/household con-sumed) set as a target where the grid is not under reach is another significant technical decision with deep political implication. The type of off-grid and mini-grid infra-structure installed determines the resulting income generation potential among small businesses, as well as the maximum number and power of the appliances owned by different households. It is on this basis that the World Bank and other agencies proposed a "multi-tier framework" to define energy supply levels (Bhatia and Angelou (2015). Figure 5.3 report a schematic framework of the means of electrifi-cation and their possible uses produced by the *International Energy Agency*.

Off-grid solutions include in fact a large spectrum of options, and it is arguable that their scale (power output) and usability (e.g. whether or not devices include a storage unit to use power overnight) define the impact they can have on development objectives. The results of the least-cost electrification in Chap. 4 showed that in order to achieve electrification tiers between 423-160 kWh/person/year and 160-44 kWh/person/year in urban and rural areas, respectively, 41% of the total capacity additions would be represented by mini-grids (mostly PV-powered) and 8.5% by standalone systems. Such figures give an idea of the relevance that decentralised solutions could have in the context of EA, and thus of why proper supporting policy is necessary to enable their deployment.

Last, concerning storage, technical and economic questions of cost shifting are deemed relevant. While mini-grid and off-grid solutions are in fact in many cases already the most cost-effective way to provide energy access to rural populations in EA countries, storage solutions that could enable round-the-clock power availability are still lagging behind in terms of their penetration. As referred by Bart Boesmans, chief technical officer for ENGIE Africa (a French energy company with a longstanding presence in the continent), *"based on the current costs of the storage technologies, it would be preferable in the short term to focus on hybrid power projects and minigrids"* (Renewable Energy World 2017). Solar generation in tandem with gas turbines offers a possible solution: if extra power is needed for

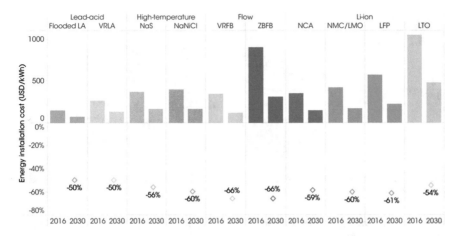

Fig. 5.4 Battery electricity storage systems, installed energy cost reduction potential (2016–2030). Source: IRENA (2017)

only a fraction of the day, the business case changes significantly. Furthermore, it has been argued that utilities making a business plan to develop a fossil fuel-based generation solution prospected to come on-line in 5-year should compare their business costs with that of energy storage solutions over 5 years to assess the economic sustainability of the project. This remarks the significance of long-term decisions in energy planning, with PV already competitive with fossil-fired generation and storage costs being the only effective determining factor. Figure 5.4 depicts, for different battery solutions (including new and emerging technologies), the current cost and the forecasted potential for cost reduction by 2030. IRENA (2017) remarks that the cost of Li-ion batteries for transport application has fallen by a striking 73% between 2010 and 2016 for and that in some high-income countries the cost of stationary applications of such storage option has also dropped significantly. The authors of the report forecast that the cost of such batteries could decrease by a further 54–61% by 2030, owing to both economies of scale and technology improvements across the manufacturing value chain. Furthermore, flow batteries, which have the benefit of being able to independently scale their energy and power characteristics, could also offer large cost reduction potential.

In addition, hydropower could also play an enabling role for the increased penetration of renewable by serving as a clean storage solution. This could be the case through the development of pumped-storage facilities, which store potential from other variable RE (such as solar and wind) by using excess power to pump water in a reservoir upstream, which can be later released down to a dam when peak power is required. As of 2018, only four pumped-storage facilities are under operation throughout SSA, all in South Africa, while according to the International Hydropower Association two have been announced in the Kingdom of Lesotho. In the future the technology (currently the only mature and largely adopted utility-scale energy storage option) could gain increasing relevance in EA, also thanks to the

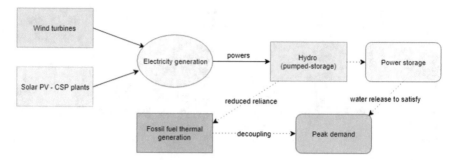

Fig. 5.5 The potential role of hydropower in promoting RE penetration and emissions mitigation.
Source: Authors' elaboration

possibility of transforming existing reservoirs to pumped hydropower schemes
(Fitzgerald et al. 2012). As evidenced in Fig. 5.5, pumped-storage might then lead
to a decoupling of fossil-generation to satisfy peak demand, and thus to a mitigation
of potential GHG emissions.

For instance, Murage and Anderson (2014) propose the cooperation between the
large Lake Turkana wind farm and a hydro-pumped storage facility in Kenya. The
authors argue that diurnal wind patterns in the region exhibit negative correlations
with load patterns, and thus storage would render energy more reliable (reducing the
system's total power output shortage by 46%) and increase the expected daily
revenue of the wind farm by 10,000+ USD. In Mekmuangthong and
Premrudeepreechacharn (2015) the combined operation of a PV plant and a hydro-
pumped storage facility is analysed, and the authors find that operating the system
across an inter-seasonal cycle results in energy and economic gains.

Furthermore, as discussed by Castronuovo et al. (2014), the introduction of
intermittent RE (such as wind or solar energy) in isolated systems can be more
complex than it is in large interconnected systems, because VRE plants are generally
unable to assist in maintaining the frequency and voltage of the system within tight
margins. Thus, also in this case storage devices can perform an important role in
balancing the amounts of power and energy in the system. Results from different
studies, for instance on the integration of isolated wind-PV systems and pumped
storage, or in hybrid diesel-RE systems, show that the implementation of storage
facilities considerably improves the scope of serving the energy needs of remote
communities (Abbey and Joos 2009).

However, it must be noted that pumped-hydropower is a mature technology with
site-specific cost (IRENA 2012) and thus there is little potential to structurally
reduce the total cost from a technology perspective. Moreover, pumped storage
involves large infrastructure and is not as scalable and modular as some of the
emerging battery electricity storage technologies.

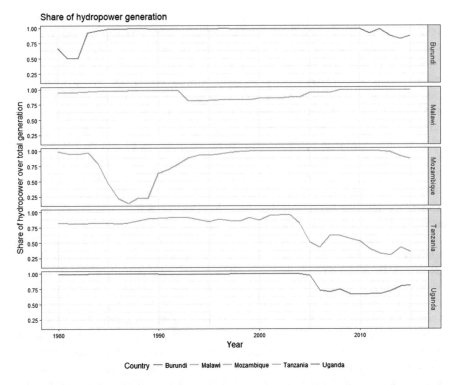

Fig. 5.6 Share of hydropower generation over total generation in selected EA countries. Source: Authors' elaboration on US EIA data (2017)

5.1.3   Hydropower Dependency and Climate Impact

Another crucial technical consideration in the context of EA concerns the long-lasting situation of hydropower dependency found in Burundi, Malawi, Mozambique, Tanzania and Uganda (see Fig. 5.6 reporting the evolution of the share of hydropower generation over total generation). These countries are currently vulnerable to water level fluctuations and future permanent changes in precipitation patterns and potential evapotranspiration levels. An increasing number of disruptions has already been witnessed over recent years due to the impact of different emerging stressors, all of which are expected to gain further significance. These include a rapidly and steeply growing population, coupled with an increase in per-capita water consumption levels associated with economic growth. Together with climate risks, these developments might jeopardise the energy security if no diversification is accomplished. To mention some examples, Malawi's hydropower capacity factor has fallen below 50% in 2015 and 2016 (authors' elaboration on IRENA 2017), with a generation of 1176 GWh in 2015, well below the 1800 GWh of 2010, despite a 18% increase in installed capacity over the same period. The water level of Lake Malawi has been in fact reported to be in a continuous negative trend since 2008 (USDA satellite altimetry

2018). Tanzania's hydropower output has also been far from constant, and the national power sector has faced severe disruptions in recent years, the worst of which was the drought of October 2015, which led to the idling of the entire hydropower capacity of the country. These fluctuations and the role of extreme events are identified in the literature as having substantial adverse effects on both total factor and labour productivity of small and medium enterprises (Conway et al. 2015; Occhiali 2016), as well as on overall development. Concerning predictions for the future of hydropower, scientific works on the expected long-lived impacts of climate change on the average hydropower availability finds mixed and highly uncertain evidence for EA (Turner et al. 2017). Nonetheless, a high level of confidence is instead expressed over the fact that climate change will heavily skew seasonal and decadal precipitation patterns, with an expected rise in the frequency and intensity of extremes (including hydrological droughts and floods), as well as increasing temperatures and thus higher levels of potential evapotranspiration (IPCC 2015).

In this setting of uncertainty, diversifying the power mix and exploiting the untapped potential of different RE resources such as solar and wind power, as well as cleaner fuels like NG where this is available and economically viable, is pivotal. Hitherto only Tanzania seems to have effectively undertaken a consistent process of diversification of its power generation mix. Moreover, investing in building hydro infrastructure that is resilient and adaptable to different water availability level and power demand scenarios is of great importance. For instance, installing multiple turbines of different sizes in new hydropower plants could allow dams to operate more efficiently over a wider range of discharge quantities, thus improving the downstream flow regime, minimising generation reduction and thus revenue loss, and guaranteeing more resiliency with respect to the uncertain future hydrology (Fig. 5.7).

5.2   Economic Considerations

With regards to the economic dimension, we touch upon three fundamental questions:

- The first concerns the on-grid/mini (off)-grid trade-off faced by public decision makers in their investment or subsidisation decisions. The concrete question is where the economically optimal geographic boundary between the two alternative solutions lies across different contexts and locations.
- The second regards the *status quo* of demand being constrained by a limited supply in many regions of EA, due both to a growing population and to the fact that ability-to-pay and willingness-to-pay for electricity services are systematically lower than market prices for most off-grid appliances (Grimm et al. 2017).
- The third point follows an argument put forward by Deichmann et al. (2011), i.e. *"how the configuration of cost-effective energy supply options will change in*

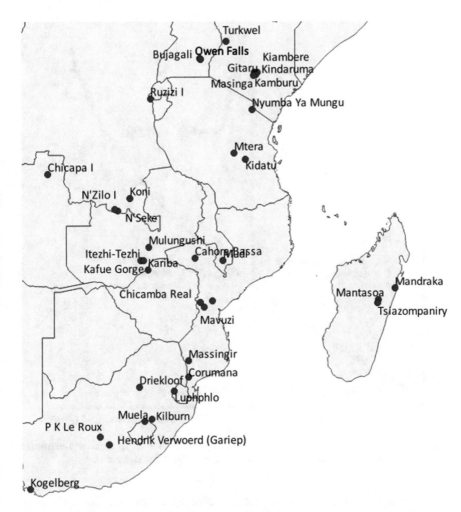

Fig. 5.7 Hydropower dams in East Africa. Source: Authors' elaboration on FAO data (2016)

the future as technical change lowers the cost of RE sources, or as premium values for clean technology change relative fuel prices".

5.2.1   On-Grid/Decentralised Solutions Trade-Off

With reference to the first issue, Fig. 5.8 presents a schematic representation of the optimisation problem faced by public decision-makers. Off and mini-grid RE-based solutions only become competitive where the distance from the grid is such that an expansion of the existing grid is inefficient. The levelized cost of electricity is a convenient metric for comparing the two alternatives, with distance-cost curves

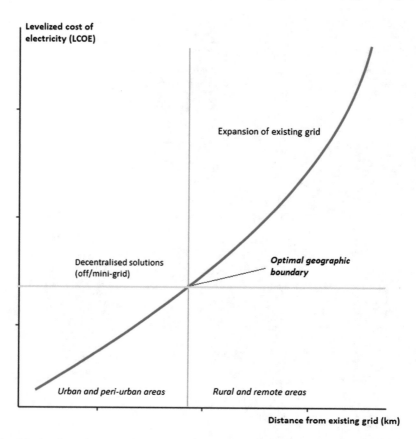

Fig. 5.8 A schematic economic framework to compare the trade-off of main grid expansion vis-à-vis off-grid solution to increase electrification. Source: Authors' elaboration

plotted according to idealised cost profiles, reflecting a commonly lived situation, rather than any geographically-determined setting (Deichmann et al. 2011). The curves could indeed have different shapes in some settings, but this idealised framework is used to formalise a very frequent scenario with which policymakers in EA have to deal. The marginal cost of grid expansion slopes upward, because new fixed investments in hard infrastructure are reaching progressively fewer and fewer consumers as the system expands in remote areas with lower population densities, more complex terrains and greater institutional constraints (IEA 2017a). In first approximation, the least-cost mix of centralized and decentralized power can be thought to depend on the cost of grid expansion, which is determined among other things by geography (Parshall et al. 2009), and by the relative cost of locally available energy sources (Mentis et al. 2017).

Estimated unit costs of grid extension in Sub-Saharan Africa range between $6340/km in densely populated areas to $19,070/km in remote rural provinces (UNIDO 2017). For instance, on average it costs Tanzania's national utility

TANESCO $2300 to extend the main grid to a rural off-grid household due to the high cost of running high voltage lines out to remote areas (standing at over $30,000/ km for >66 kV transmission lines). Such high costs—usually borne by Governments—also have the side-effect of preventing investments in new transmission capacity. Conversely, the mini-grids could bring access to the same household for $500—$1000 as a result of cutting out high voltage lines. While access with on-grid and mini-grid is not equivalent, electricity provision *per se*, even at low levels of per-capita consumption, can pay a substantial role in breaking energy-poverty traps and fostering development, employment and poverty reduction (see empirical evidence in Dinkelman 2011; Khandker et al. 2013). In this sense, bridging private capital would also reduce the burden on the national governments, which would incur in less subsidy payments. According to IRENA (2016), the installation cost of stand-alone solar PV mini-grids in Africa is as low as $1.90/W for systems larger than 200 kW. At the same time, solar home systems provide the annual electricity needs of off-grid households for as little as $56/year. Besides marginal costs, there also exists an issue of grid resilience. Mini-grids with battery storage and local distributed generation are in fact typically more resilient than those that rely on extensive transmission lines, especially through forested landscapes.

According to IEA's (2017b) *New Policies Scenario*,[2] main grid extension would serve half of all newly connected households in the entire SSA by 2030. This result is highly consistent with the results of the scenario analysis carried out in Chap. 4, which results in a median share of grid capacity additions of 50%. Always according to the IEA, in rural areas decentralised power systems would instead be the most cost-effective solution for more than 65% of those who will gain access. Previous multi-scenario spatial optimisation analysis for SSA [such as in Deichmann et al. (2011) and Mentis et al. (2017)] seem to suggest that—above very low consumption levels—decentralised solutions would be the lowest cost option only for a minority of households in SSA, i.e. that the optimal geographic boundary is rather remote from the main grid, even when potential future cost reductions of decentralised solutions and the costs of transmission and distribution infrastructure of grid-based electrification are considered. While our results agree on the modest share of standalone PV and diesel solutions, we find that mini-grids (and in particular PV) are a viable way to provide access with medium tiers of electricity consumptions which allow to avoid mass investment to extend the national transmission grid, at least in the next 10–15 years. Previous stories of mass electrification, such as the case of India, where over 99% of those who have gained access since 2000 have achieved it as a result of grid extension, are interesting in comparative terms. However, it must be noted that the mean population density of India is much higher than that of EA, allowing grid expansion to target on average a higher number of new potential

---

[2]The New Policies Scenario *"takes account of broad policy commitments and plans that have been announced by countries, including national pledges to reduce greenhouse-gas emissions and plans to phase out fossil-energy subsidies, even if the measures to implement these commitments have yet to be identified or announced"* (IEA 2017a).

consumers. Conversely, it also seems that there has been no universally effective strategy for boosting electrification rates in recent years, reflecting the broad diversity of energy resources, infrastructure, institutions, population density and distribution, and geography across regions and countries in the world.

The need to set the national and regional fuel mix on a low-carbon trajectory, especially in sight of the steep regional population growth, makes large-scale grid connected renewable projects such as solar thermal power and large wind farms particularly relevant. Note that from the household perspective the two alternatives (on-grid access and off-grid back-up solutions) may well be seen as complements rather than mere substitutes, since individuals and small firms can be reluctant to rely exclusively on poorly managed central utilities with interrupted-supply issues, or because they may simply want to hedge subsistence economic activities and small enterprises from outages or volatile prices. Overall, from a policy perspective, on-grid investments could initially be concentrated on certain prospective regions with high business potentials or industrial zones to which firms might relocate, while mini and off-grid technology could be used as a complement to promote energy security and development means to more remote areas. The limited amount of energy that off-grid technology is currently capable of producing renders it challenging to serve an array of different energy needs (including industrial uses), and it requires additional capital investments e.g. for storage. Thus, an integrated on-grid/off-grid mixed strategy would enable the development of industrial and service sectors, and at the same time achieve broad household and rural enterprises access to electricity at relatively low cost.

5.2.2 Household Ability/Willingness-To-Pay for Electrification

With regards to the demand-side issues of electricity markets, the numerical results of a recent RCT-based study for off-grid PV appliances in rural Rwanda carried out by Grimm et al. (2017) provide relevant insights. The field experiment found that estimated willingness-to-pay (WTP) for different typologies of appliances in villages where the grid is missing are all clearly below the respective market prices, with elicited values ranging between 38 and 52% of the latter. Despite being only a case-study with a limited external validity, the experiment highlights that if electrification efforts are to be successful in remote rural areas of EA, an approach based on private market alone is unlikely to reach the broader population. Off-grid technology requires public subsidies, otherwise in all likelihood there will be a lack of demand at market prices, also because the alternative and currently predominant option of traditional biomass has no explicit price or cost to anchor on. Thus, it seems that offering a publicly-subsidised but privately-purchased alternative option to deliver off-grid electricity at the household level could be a way to render biomass collection and the related social costs at least more visible, and thus to reduce energy-derived

Table 5.2 Current retail price of grid electricity for domestic users

Country	Price range (USD/kWh)
Burundi	0.046–0.37
Kenya	0.23
Malawi	0.096
Mozambique	0.094–0.14
Rwanda	0.12–0.20
Tanzania	0.046–0.18
Uganda	0.04–0.19
Average	0.125

Sources: TANESCO (2016), AllAfrica (2017), IWACU Burundi (2017), The New Times (2017), Nyasa Times (2018), Stima Consultancy (2018) and Techjaja (2018)

externalities. While fundamental needs such as cooking are not easily and cheaply satisfied with off-grid electricity, the latter offers a chance for more productive uses of energy and thus a long-run prospect of economic development. Note that while public subsidies are of great importance, it has been argued (Moreno and Bareisaite 2015) that smarter payment schemes with longer payment periods, for example in the form of mobile-phone-based 'pay-as-you-go' plans, could also enable household investment into off-grid technology. A similar issue is also faced in areas that are within reach of the grid, but where households are credit-constrained and thus struggle to afford the high upfront connection costs, which often are well above their subsistence income. Thus, payment tools and policy can have a determinant role in unlocking households' access in both off and on-grid settings (Table 5.2).

5.2.3 Uncertainty Over Future Costs and Developments

The third question on the economics of RE deployment is the most problematic to address, because it concerns the dynamic nature of electrification investments and renewables infrastructure expansion, and in particular the uncertainty about future cost profile shifts. The issue has a twofold character: on the one hand, it refers to how technical change will affect the cost of RE sources and storage solutions. On the other, it is affected by how 'premium values' for clean technology will change relatively fuel prices, e.g. as the result of the introduction of a global carbon tax or due to shifting consumer preferences. Overall, this point concerns the question of how and to which extent the configuration of cost-effective energy supply options will differ from the current situation. Figure 5.9, drawn from IRENA's (2018) *Renewable Power Generation Costs* report, illustrates the change in the global levelized cost of electricity from utility-scale renewable power generation technologies between 2010 and 2017. The diameter of each circle represents the size of the project, while the dark band represents the fossil fuel-fired power generation cost range.

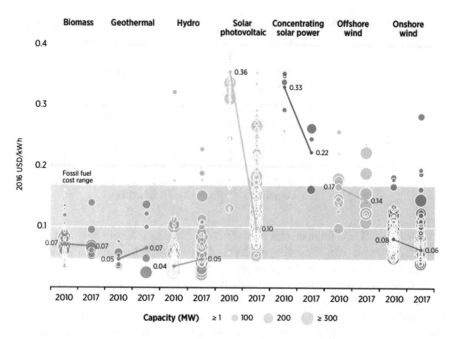

Fig. 5.9 Levelized cost of electricity (2010–2017 trend). Source: IRENA (2018)

While the LCOE figures stem from a global-scale assessment, they provide a hint of the underlying global trends in the economics of power generation. In particular, they show that while fossil-fired generation is assumed to have remained at a steady LCOE, between 0.1 and 0.2 USD/kWh, solar PV and CSP projects have witnessed a steep fall in the average LCOE with a ~70% decline for PV and a ~30% drop for CSP. However, while the decision of developing a PV field now seems competitive with that of a fossil-fired plant, the current LCOE cost of CSP does not seem to be in the competitive range. Wind power also shows a declining cost trend, and now both onshore and offshore projects look like competitors of thermal generation. On the other hand, less significant shifts are witnessed for other RE options, with biomass, geothermal, and hydropower projects not having substantially altered their cost component over the last years. Overall, it is of utmost importance to consider the trends of the LCOE lines in the planning horizon of power generation projects in EA. Technology lock-ins and path dependencies should in fact be minimised with the adoption of resilient electrification plans capable of performing efficiently in an array of scenarios of future costs of energy and technology.

5.3 Transboundary Cooperation

5.3.1 Infrastructure

In the context of expanding energy access in EA and promoting a sustainable (energy) development in the region, it is also worth discussing the role that strategic cross-country cooperation can play. Insufficient power generation is indeed not the only determinant of the current low supply and access levels in EA: the scarcity of interconnectors to trade energy resources and power across border is another leading explanation. Joint capacity additions, interconnected grids, and resource sharing could all play a major role in the achievement of a faster and more inclusive energy development and electrification process in EA. A range of bilateral agreements for power exchange already exists between neighbouring countries in EA, including the following grid connections (existing, currently under construction, or approved): a 400 kV South Africa-Zimbabwe line through Botswana (and, as of 2018, a plan to establish a second transmission interconnector boosting the wheeling capacity by 1500 MW); a 535 kV HVDC line between Mozambique and South Africa (used to export power produced at Cahora Bassa hydropower plant); a planned 400 kV line between Mozambique and Zimbabwe; two 400 kV linking South Africa to Maputo via Swaziland to power an aluminium smelter in Mozambique; a 110 kV mini-power pool formed by eastern DRC, Rwanda and Burundi managed by SINELAC; a 330/400 kV Zambia-Tanzania-Kenya-Ethiopia interconnection (with a memorandum of understanding signed in 2014 and the project, with expected completion date in 2021, currently at different phases of development in the three countries involved); a 132 kV Uganda-Kenya connection (set up in 1955), to be expanded to 300 MW in 2020 and 600 MW in 2025; a Rwanda—Tanzania interconnection project with capacity of 200 MW by 2020 and 1000 MW by 2025; a bi-directional 500 kV HVDC (2000 MW) Kenya-Ethiopia, with construction works started in 2016; Rwanda-Uganda are to be connected at level of border towns in 30 kV and 220 kV (general interconnector under construction); a 400 kV Mozambique-Malawi line with feasibility study phase completed; furthermore, the expression of interest for feasibility studies for a 400 kV interconnection between Uganda-Rwanda-Kenya was made in late 2014 (Fig. 5.10).

However, regional interconnections have not been optimized in past decades, and furthermore they have often been affected by the failure to meet contractual obligations, partially due to the lack of a contextual coordinated planning for the expansion of generation capacity in past bilateral system interconnections projects (Expogroup 2017). With a coordinated and interconnected system, it is possible to make sure that the currently cheapest power is always consumed first. In the presence of excess capacity from one source or site it is possible to complement others by selling excess supply beyond the national borders and producing a win-win scenario. Furthermore, grid interconnections increase the market size, reinforcing economies of scale. Thus, interconnection of infrastructure can reduce price and supply risks and increase the attractiveness of the local market for foreign investors in the energy sector.

(a)   *(b)*

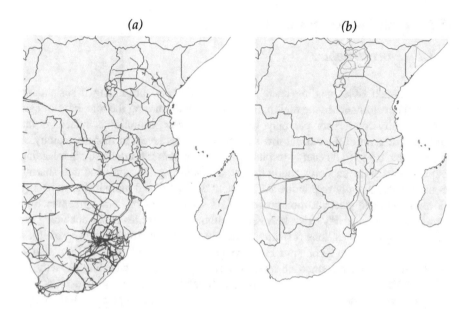

Fig. 5.10 Existing (**a**) and planned (**b**) transmission grid in East Africa. Source: Authors' elaboration on Arderne-World Bank (2017)

Interconnections can also enable countries to optimize domestic energy resources and counterbalance seasonal or natural variability of renewables. Additionally, EA countries endowed with substantial gas resources, namely Tanzania and Mozambique, could increase their exports of either gas or power (or both) in the region, contributing to complement fossil resources scarcity and to cope with RE intermittency problem, while also boosting economic cooperation in other sectors through multi-party trade agreements. As a matter of fact, endowment heterogeneity is, at least in principle, a strong engine in boosting cooperation over boundaries in the region (although it could also lead to disputes over contested borders such as in the case of Malawi and Tanzania over Lake Nyasa's fossil fuel resources). In particular, energy interdependency is likely to promote integration and could spill cooperation over into other areas, especially where permanent shared infrastructure is put into place.

5.3.2   The Eastern African Power Pool

Achieving effective energy transboundary cooperation requires political will from multiple parties and converging economic interests. As of today, efforts in the region have culminated in the establishment of the Eastern African Power Pool (EAPP) in 2005, currently comprising 11 EA countries. The EAPP represents an ambitious attempt to achieve an *"optimum development of energy resources in the region and to ease the access to electricity power supply to all people of the countries in the Eastern Africa Region through the regional power interconnections"* (official mandate). In other words, it seeks to ensure increased power supply, reduced electricity production costs, and efficient transmission and exchange to ultimately establish a modernized electric market in EA. A regional master plan is being regularly updated to support the drafting of national power generation, transmission and export plans in EA countries. For instance, the Ethiopia-Kenya grid connection currently under construction was supported by EAPP, and its capacity (500 kV HVDC) is projected to be sufficient not only for the exchange between the two countries but also for future potential interconnections to other countries in the region (e.g. Tanzania).

Nonetheless, as remarked by the Secretary General of EAPP, Mr. Lebbi Mwendavanu Kisitu Changullah (ESI Africa 2016), infrastructure financing is proving to be the main roadblock to the implementation of transboundary energy projects in EA. Furthermore, the reinforcement of the existing transmission network remains an equally important challenge. The large extent of the region and the non-homogeneity in the distribution of resources are further elements to account for in the formulation of transboundary projects in EA.

In 2012, countries in the region began the Eastern Africa Integration Programme, aiming at the connection between the power grids of Ethiopia, Kenya, Tanzania, Uganda, and Rwanda in three phases. The first phase of the programme, connecting Ethiopia and Kenya, is under implementation. The (Tanzania-Zambia) TAZA project constitutes part of the second phase, with the other part—the Kenya-Tanzania transmission line—already under construction. In July 2018 Kenya and Tanzania secured more than $600 million in funding from international financiers for large-scale interconnection and power pooling (The East African 2018). The ultimate objective is the accomplishment of a regional Transmission Corridor Development project that will see Tanzania link the East Africa Power Pool to the Southern Africa Power Pool. This would guarantee the existence of a large competitive power market allowing to meet the energy security needs of the region in a cost-effective manner. The ability to engage in short-term trade, either bilateral or through existing market mechanisms in SAPP, will further enable countries to diversify their energy mix, eliminating the need for expensive emergency power during supply shocks, and improving conditions for the development of scale-efficient generation infrastructure selling to regional power markets. In this context, Tanzania would have a key role for its strategic position between the East and the South of the continent. For this reason, the country plans to boost power generation capacity from the current 1500 MW to 5000 MW over the next 3 years by building new gas-fired and hydroelectric plants, according to the country's Energy Ministry.

Table 5.3 Organisations, institutions and frameworks involved with energy cooperation in EA

Name	Countries involved
Eastern Africa Power Pool (EAPP)	Burundi, Democratic Republic of Congo (DRC), Egypt, Ethiopia, Kenya, Rwanda and Sudan
Energy and Environment Partnership for Southern and East Africa (EEA)	Botswana, Burundi, Kenya, Lesotho, Malawi, Mozambique, Namibia, Rwanda, Seychelles, South Africa, Swaziland, Tanzania, Uganda, Zambia and Zimbabwe
East African Community (EAC)	Burundi, Kenya, Rwanda, South Sudan, Tanzania, and Uganda
East African Centre for Renewable Energy and Energy Efficiency (EACREEE)	Burundi, Kenya, Rwanda, Tanzania, and Uganda

5.3.3 Energy Resources Sharing and Water Basins Management

The sound management of transboundary RE (such as hydropower) and of fossil resources found near countries' borders should underpin all future infrastructure projects. The EAPP, the EEP (*Energy and Environment Partnership*) and the other institutions and arenas where regional energy coordination takes place should constantly review the various players involved in the strategic energy planning across the region and update a plan that embeds current and future projects that have been proposed or already under consideration (Kammen et al. 2015). This would increase the overall efficiency of investments. Table 5.3 reports a list of currently operating organisations, institutions and frameworks devoted to energy cooperation in EA.

It must also be noted that given the current situation of hydropower dependency and future capacity expansion plans, improving transboundary water resources management and within-country efficiency in water use among competing sectors is essential. A large number of transboundary river basins stream through EA countries (Fig. 5.11), calling for tight coordination to achieve optimal equilibria (Namara and Giordano 2017), since water availability downstream (and water infrastructure management) is largely affected by political and infrastructural choices upstream (Grey et al. 2016). Cooperative governance could in fact reduce water conflicts (and their spillage in other arenas including the power sector), increase efficiency in resource use—including hydropower output—and create added economic value by internalising potential negative externalities from uncoordinated action, boost investment and financing of shared water infrastructure (such as Pareto-efficiently located dams). The relationship between hydropower and rural irrigation in multipurpose reservoirs is also pivotal.

Operations limited to individual dams are in fact prone to be insufficient to mitigate region-wide, river basins, and reservoir-level impacts, and balancing hydropower and other uses is likely to be most effectively achievable at wider scales. Thus, shifting away from single-project focus and leaning towards basin-level/

Fig. 5.11 River basins in
EA. Source: Transboundary
Freshwater Dispute
Database (2017)

regional and integrated approaches is suggested. Currently a complex regime exists in the regulation and agreements of transboundary watercourses. Table 5.4 reports a list of the local institutions (thus excluding international organisations and foreign authorities) active in different water resources management in EA. Figure 5.11 illustrates the location and areal extent of the major river basins in the region.

5.4   Key Policy Challenges

There is extensive evidence that a high share of the energy projects commissioned in SSA countries (and in particular those involving RE) have witnessed failure, delay in delivery, or cost skyrocketing (Ikejemba et al. 2017). This is certainly a relevant consideration from which to start the discussion of the most fundamental policy aspects to be considered by public and private stakeholders involved in the energy sector of EA countries. Supporting policy for potential deployment is in fact of

Table 5.4 Transboundary river basin regulatory authorities

Name	Countries involved	Mission and objectives
Zambezi River Authority	Zambia, Zimbabwe	Operate and maintain the Kariba Dam
Zambezi Water-course Commission	Angola, Botswana, Malawi, Mozambique, Namibia, Tanzania, Zambia and Zimbabwe	To promote the equitable and reasonable utilization of the water resources of the Zambezi Water-course as well as the efficient management and sustainable development thereof
African Ministers' Council on Water (AMCOW)	In EA and surrounding regions: Burundi, Democratic Republic of the Congo, Mozambique, Rwanda, South Africa, Uganda, Tanzania, Zimbabwe	Develop enabling frameworks, strengthen collaboration with civil society and partner institutions, review, monitoring and reporting, governance, capacity building and training
Lake Tanganyika Authority (LTA)	Burundi, DR Congo, Tanzania, Zambia	Lake Tanganyika Authority supports, coordinates, monitors and evaluates the implementation of the Convention on the Sustainable Management of Lake Tanganyika. It also oversees the implementation of program and project activities
Lake Victoria Basin Commission (LVBC)	Kenya, Uganda, Tanzania	Coordinates the various interventions on the lake and its basin and serves as a centre for promotion of investments and information sharing among the various stakeholders. Activities focus on harmonization of policies and laws, monitoring, management and conservation of aquatic resources, development of economic activities and infrastructure
Ruvuma Joint Water Commission	Malawi, Mozambique, Tanzania	
Nile Basin Initiative (NBI)	Among EA countries: Burundi, DR Congo, Kenya, Rwanda, Uganda, Tanzania	Identifying and preparing investment projects for the development of shared water resources; facilitating agreements between countries for investment financing and for future management through the national agencies (irrigation and drainage, watershed management, flood early warning and protection, fisheries, power interconnectors and power generation)

(continued)

Table 5.4 (continued)

Name	Countries involved	Mission and objectives
Inco-Maputo Tripartite Permanent Technical Committee	Mozambique, South Africa, Swaziland	Management of the water flow of the Inkomati River and Maputo River specifically during times of drought and flood; recommendation of measures to protect and develop these water resources
Komati Basin Water Authority	Mozambique, South Africa, Swaziland	Implement the Komati River Basin Development Project focused on hydropower development (construction, operation and maintenance of the Driekoppies Dam and Magauga Dam)
Limpopo Watercourse Commission (LIMCOM)	Botswana, Mozambique, South Africa, Zimbabwe	technical advice on matters related to the development, utilisation and construction of water resources in the Limpopo River basin

particular importance for the successful exploitation of RE. Irrespective of the heterogeneous resource endowment, key issues have similarly affected different countries, including the political agenda; the process of awarding projects to public and private companies; the financing mechanisms; stakeholder dialogue and co-operation; the planning and implementation dynamics; maintenance programs; as well as public inclusion. In this context, we review the current policies supporting energy investment and RE deployment in the EA countries under examination, while also discussing the key lessons learned and the policy challenges faced by public policymakers but mainly affecting private actors.

5.4.1 Competition, Investment Attractiveness, and the Role of IPPs

It has been extensively discussed that in the task of assigning large energy infrastructure projects, competitive bidding promotes efficiency (Ackah et al. 2017). Nonetheless, until recently most SSA countries have used mostly non-competitive direct negotiation to procure additional generation capacity (Table 5.5). While in principle this could be perceived by public decision makers as a quicker option to procure infrastructure development, it has large potential negative effects. It is in fact prone to increase the final user cost of energy (since it does not incentivise cost-efficient solution), deter further private investment due to burdensome processes (where private and personal relationships affect public choices), and promote the allocation of licenses to companies that may not have sufficient capacity to deliver. On the other hand, countries that build tendering capacity and guarantee an open and

Table 5.5 Level of competition and unbundling in the grid/off-grid power sectors

Country/Source	On-grid	Decentralised solutions
Burundi		
Kenya		
Malawi		
Mozambique		
Rwanda		
Tanzania		
Uganda		
South Africa		

Source: Climatescope (2017); various governmental websites
Red: No competitive market; *yellow*: Measures to pursue competition in the market have been implemented; *green*: Competitive market in place

competitive bidding process are more likely to attract large foreign and institutional investors with good financial and technical capacity.

Concerning this latter point, Eberhard et al. (2017) characterised the growing contribution of IPPs in SSA by their reliance on long-term contracts with off-takers. In fact, development finance institutions such as the World Bank and the African Development Bank are playing an increasingly important role in attracting private sector interest, financing IPPs, and mitigating risk, in particularly due to their capacity to influence governments to honour commitments. However, for private investment in the energy sector to flourish, EA needs dynamic planning, linked to competitive procurement of new generation capacity. This must be accompanied by the building of regulatory capacity that encourages the distribution utilities that purchase power to improve their performance and prospects for financial sustainability. In particular, critical success factor for IPPs includes a range of both country-specific and project-specific factors. Furthermore, the presence of supportive policies during and after project implementation has proved to be very effective in guaranteeing returns on energy investments, with research and development playing a major role (Mas'ud et al. 2016). Note that not only local but also global policy setting is a determinant of success, because it has a direct impact on industrial and R&D sectors while also determining transfers of experience and knowledge (Pillot et al. 2017).

The relatively young market for off-grid and mini-grid decentralised solutions has emerged as a much livelier and more competitive field than that for grid capacity additions and grid extension. The SA/MG market is indeed being populated by a growing spectrum of private companies, since it does not present the hard-fixed costs necessary to expand on-grid generation, which can often only be borne by Government utilities or large international developers. Thus, it has also been much more dynamic in terms of the pace of its growth and the degree of penetration, and it is one of the reasons why projections (Chap. 4 of this book; IEA 2017a, b; PBL 2017) suggest a relatively high degree of mini-grid development in EA in the coming year, with the rise of a coexistence between centralised grids and not interconnected MG and SA systems.

Another policy measure that is proving successful in terms of the objective of increasing competition in the energy market is the separation of the management of state-owned power generation utilities (which in EA countries are all public and have constantly been in economic deficit) and the authorities responsible for the main transmission system. This aims at providing a level-playing field for IPPs and at increasing investment confidence. This separation has indeed already been implemented in Kenya or in Uganda, and it is taking place in Malawi. Table 5.6 reports the main authorities (including those responsible for generation, transmission, regulation, rural development and policymaking) involved in the power generation market of each EA country.

In Burundi there is no competition in the on-grid market, and the power sector is vertically integrated. REGIDESO, the utility responsible for electricity generation and distribution, is entirely owned by the state. This makes sense, given the small size of the country, the very limited extent of the transmission (546 km) and distribution (337 km) grids and the relatively limited generation (263 GWh). Most of the potential to increase the currently extremely low electricity access rate (10%) lies with decentralised solutions, mostly because the bulk of the population (87%) lives in rural areas away from the grid. The country has a fairly developed legal and policy framework for mini-grids (updated in 2015), which allows private ownership and operation and establishes a duty exemption for PV array and modules and power generators. Although no mini-grids are yet in place, a project of 7 first mini grids in the Solar Electricity service with Mini Grids in Africa-Burundi (SESMA-Burundi) is at the feasibility study stage.

In Kenya, a monopsony characterises the energy sector structure, with a single buyer (the Kenya Power and Lighting Corporation, KPLC) interacting with many would-be sellers (IPPs), determining a partially vertically unbundled energy market. In 2006 an ad-hoc electricity regulator was established (Energy Regulatory Commission), while in 2008 KETRACO (the Kenya Electricity Transmission Company) was founded as the transmission system operator, taking away such responsibility from KPLC. Private sector participation is found in generation (owning at least 10% of the total installed capacity and with at least six IPPs active in the country), while a bill was passed in 2015 (the Kenya Energy Bill) to establish a distribution licensee plan enabling any person in the licensee's area of supply to receive a supply of electrical energy either directly from the licensee or from an accordingly authorised electricity retailer. With regards to the decentralised solutions market, mini-grid have been regulated since 2012 with licensing, stable-tariffs guaranteed and clear rules on interconnection. While no subsidies exist, duty exemptions are in place for power generators, energy storage systems and monitoring systems, and specific market financing facilities are available to support operators to develop mini-grid and stand-alone systems.

In Malawi the power sector has long been characterised by a vertically integrated monopolistic structure, with ESCOM being responsible of generation, transmission, and distribution. In late 2016, the government completed the unbundling of the

Table 5.6 Authorities in the energy market of EA countries

Authority	Burundi	Kenya	Malawi	Mozambique	Rwanda	Tanzania	Uganda	South Africa
Energy regulator	Agency for Control and Regulation	Energy Regulatory Commission	Malawi Energy Regulatory Authority (MERA)	Energy Regulatory Authority (ARENE)	Rwanda Utilities Regulation Authority	Energy and Water Utilities Regulatory Authority	Electricity Regulatory Authority	National Energy Regulator of South Africa (NERSA)
Power utilities	Water & electricity Utility for Production and Distribution (REGIDESO)	Kenya Electricity Generating Company; Kenya Electricity Transmission Company; Kenya Power and Lighting Company	Electricity Supply Corporation of Malawi Limited (ESCOM); Electricity Generation Company (EGENCO)	Electricidade de Mocambique (EDM)	Rwanda Energy Group; Energy Development Corporation Limited; Energy Utility Corporation Limited	TANESCO (Tanzania Electricity Supply Company Limited)	Uganda Electricity Generation Company; Uganda Transmission Electricity Company Limited	ESKOM
Rural electrification agency	Rural electrification agency for Burundi	Rural Electrification Authority	Malawi Rural Electrification Programme (MAREP) under ESCOM	National Fund for Rural Electrification (FUNAE)	Rwanda Energy Group/Energy Utility Corporation Limited	Rural Energy Agency	Rural Electrification Agency	–
Ministry	Ministry of Energy and Mines	Ministry of Energy and Petroleum	Ministry of Natural Resources, Energy and Environment	Ministry of Energy	Ministry of Infrastructure	Ministry of Energy and Minerals	Ministry of Energy and Mineral Development	Department Of Energy

power sector, taking away from ESCOM the responsibility to manage power generation and establishing EGENCO to serve this purpose (Millenium Challenge Corporation 2017). Note that currently Malawi is not connected to the power systems of neighbouring countries and cannot therefore engage in power trading. Since 2007 an independent electricity regulator (the Malawi Energy Regulatory Authority) is controlling the power sector. Despite the recent partial vertical unbundling, competition in the generation sector is still very weak, although in 2017 the country has started opening the electricity market with a standardized power purchase agreement to allow IPPs to operate. Several IPPs are currently developing projects and more than 36 MoUs have been signed so far, the majority for RE projects. Mini-grids have been regulated since 2004 (Rural Electrification Act), they have a dedicated regulator and a dedicated team within the national utilities. Duty exemptions are provided for in the Customs and Excise Act 2014, and standardised PPAs with partially cost-reflective tariffs have been put in place to support the develop of stand-alone systems, while clear rules on their eventual interconnection have also been established.

Mozambique is also characterised by a monopsony situation, with horizontal unbundling having interested only the generation segment, and the main player being the state-owned Electricidade de Moçambique (EDM), responsible for the generation, transmission, distribution and sale of electricity. EDM is however controlling only 20% of the country's capacity, with the bulk of it coming from IPPs (after the publication of the Public-Private Partnership law in 2011) and, chiefly, from Cahora Bassa Hydro (owned by ESKOM South Africa, the Governments of Mozambique and Portugal). With regards to IPPs, the 2011 bill foresees that all of them must sell electricity to EDM and negotiate prices on a contract-by-contract basis, thus determining the monopsony. Currently, four IPP plants are operating in Mozambique, accounting for around 500 MW. Decentralised solutions are supported by FUNAE, the National Fund for Rural Electrification and they have been regulated since 1997. In 2018, CRONIMET Mining Power Solutions and MOSTE have signed a MoU with FUNAE to develop Mozambique's first privately developed and financed mini-grid (expected to generate up to 200 kWp of solar power) on Chiloane Island, which will also be the largest pre-paid solar mini-grid in the country. Upon successful implementation of the Chiloane Island mini-grid, the consortium expects to develop a portfolio of 60 or more mini-grids across Mozambique.

In Rwanda, the power sector has been recently unbundled, and a very large participation of IPPs is found in generation, which own 80% of the installed capacity. The Rwanda Electricity Group (REG), established in 2014 and entirely owned by the government, is the main electricity company. Operations are carried out by two subsidiaries—the Energy Development Corporation Limited (EDCL), which supports new capacity and transmission development both by itself and by IPPs, and the Energy Utility Corporation Limited (EUCL), operating transmission and distribution networks and selling the power. At the same time, Rwanda has one of the smallest energy sectors in SSA, with only 208 MW of installed capacity. Nonetheless, the country has an oversupply of on-grid generation and a backlog of contracted projects, and thus IPPs will not be able to secure permitting in the country until the roughly 200 MW of pipeline projects with existing PPAs are moved further into development. Currently, IPPs must go through the Rwanda Development Board (RDB) to start the process of developing new projects, before negotiating a PPA with EDCL and signing with EUCL. In the past, ~20 year "take-or-pay" PPAs have been signed, which compensate for full output of their project, and as a result have determined a situation where Rwanda has higher electricity tariffs than the EA average. However, all future IPPs must now participate in a competitive tender process, monitored by the Rwanda Utility Regulatory Authority (RURA). Moreover, the government has also introduced an array of policy instruments to render on-grid development attractive for the private sector, including tenders, unsolicited proposals and favourable tax regimes. These have drawn many private-sector players, and around 50 PPAs have been signed to date. Additionally, Rwanda has signed a 30 MW PPA with Kenya, but the transmission infrastructure has not yet been built.

In Tanzania the key player in the power sector is TANESCO, the Tanzania Electric Supply Company Limited, which owns most of the country's transmission and distribution network and more than half of its generating capacity. Currently, IPPs' capacity share stands at 19.3%. However, TANESCO has outstanding debts with IPPs for close to 0.5 billion USD, due to the excessively low electricity rates. This has resulted in holding back investment in upstream capacity. Thus, the sector is still vertically integrated, and the Electricity Supply Industry Reform Strategy and Roadmap, which foresees TANESCO to be unbundled by 2025, was passed by the Government only in 2014. On the other hand, Tanzania has one of the most robust regulatory and legal frameworks in EA, encouraging the construction of small power projects. At end-2015, the second-generation framework was approved. This establishes that projects will earn a fixed tariff for the lifetime of the standardised PPA, instead of having annually fluctuating rates based on the distribution network operator's avoided costs. The selection method will also vary by technology, with an administrative process for small hydro and biomass projects, and competitive bidding for solar and wind. At the same time, a lively environment has emerged in terms of off-grid energy providers, partially thanks to governmental subsidies for generation and storage. These includes grants for pre-investment activities, such as feasibility studies, environmental and social impact assessments, and performance grants for electricity connection, as well as duty exemptions for mini-grid systems. Tanzania is in fact a hotspot for the distribution of pico-solar lighting products and the development of mobile-based, pay-as-you-go business models for access to off-grid solar arrays.

In Uganda the power sector has been fully vertically and horizontally unbundled in 2001, in so far that there are legally separate private companies at each segment of the pre-retail power system. However, a single-buyer (the Uganda Electricity Transmission Company Limited) situation persists, irrespective of a 60% IPP capacity share. Umeme is the key responsible for distribution, with a share of 98% of total Uganda's electricity consumption. The bulk of the country's generation comes from both hydro plants owned by the state-run Uganda Electricity Generation Company (UEGCL) and by the private IPP Bujagali Energy Limited. The market is regulated by the Electricity Regulatory Authority, established in 1999. Thanks to the 2013–2014 reform of feed-in-tariffs, private companies have been playing an increasingly significant role in the national market for capacity additions, mainly in biomass and small and medium hydro. However, it must be noted that owing to grid capacity constraints and burdensome processes of grid connection, projects have often exhibited long lead times before their effective on-lining. The standalone PV market has also steadily grown over the last decade, with new players, including foreign investors, entering the market. Policy measures such as tax exemptions for equipment for solar and wind generation and subsidies for end-users have also supported expansion of the sector, thus contributing to producing a fairly competitive market.

In South Africa the public utility Eskom is responsible for 95% of the electricity consumed, determining a de-facto monopoly in the generation market. Furthermore, all the remaining small IPPs (which have begun entering the market mostly with RE projects in the framework of the auction programme which entered into force in 2011 to replace FiTs) resell their power directly to Eskom, creating a monopsony market situation. Eskom is currently also responsible for transmission and distribution, determining no unbundling of the power sector. Power off-take risk for independent generators has thus been defined risky by Climatescope (2017). In the prospect of shifting away from coal (currently accounting for the bulk of total generation), the government has agreed upon a plan of gas-RE tandem, which will necessarily imply the deployment of large-scale CSP, PV parks, and wind farms. Thus, the successful implementation of the Renewable Energy Independent Power Producer Procurement (REIPPP) Programme will certainly require greater private-sector engagement and the accomplishment of a more competitive generation market over the next decades. As far as decentralised solutions are concerned, mini-grids have been regulated since 2006 by the Electricity Regulation Act, but in the South African context, characterised by a high electrification rate and by the large extension of the national grid, they have only been meaningful in very remote rural communities.

5.4.2 Subsidies, FiTs, and Policy Instruments

On the demand-side, one of the key mission for policymakers is to find the proper financial solutions to lower connection charges and expand access to grid electricity, or create supportive policy for the uptake of decentralised solutions. Golumbeanu and Barnes (2013) highlight that, from the household's perspective, potentially accessible energy (for instance as a result of an extension of the national grid or of the opportunity to install a small PV) does not automatically imply actual access (see evidence from rural Kenya by Lee et al. 2016). In the case of on-grid connections, where potential customers must decide whether to connect to the network, up-front charges are the greatest barrier, while service provision costs are much more modest. The (IEA 2011) showed that there is indeed a strong inverse linkage between connection charges and national electrification rates in Sub-Saharan Africa, determining a determination coefficient of 0.85 when controlling for GDP. Minimum reported connection charges are the following in selected EA-7 countries (Golumbeanu and Barnes 2013 and other online sources): Kenya 400 USD (for single-phase, domestic customers within 600 meters of an existing transformer), Rwanda 350 USD, Tanzania 297 USD, Uganda 125 USD (single-phase, no-pole service, tax included).

In some way, the same issue also applies to off-grid RE solutions, where purchasing and installing the devices are the only costs faced, since generation is then free of charge (besides maintenance). As a result, the question of how to lower upfront costs seems like a most compelling one for policymakers, because it represents the key barrier to the solution of a loss-loss situation for both electricity supply companies, private providers, and potential customers. Thus, questions of private-public financing (e.g. of how large subsidies should be and which schemes they should follow), as well as bureaucracy involved, supply reliability, and behavioural considerations are all to be considered. Both payment schemes and public subsidies are pivotal in this sense. There are in fact different ways to lower up-front costs for households, such as directly subsidizing some of the connection charge; incorporating part or all of the charge into the electricity tariff; financing the charge through an external bank institution; or allowing consumers to pay the connection charge over time through credit schemes provided by the utility (including new instantaneous payment solutions enabled by digital technologies). Bernard and Torero (2015) estimated the effect of distributing discount vouchers on the demand for connections in Ethiopia, and found that the demand is very responsive, remarking that connection fees represent a very significant barrier to electricity adoption.

In each case subsidies should however be designed so as to improve access to electricity without distorting energy markets and bringing about inefficiencies. This requires them to be set-up *ad hoc* depending on the economic, geographic, and institutional context, as they imply substantial public costs. Success stories of electrification backed up by different typologies of subsidies in Brazil, China, India, Thailand, Costa Rica and Tunisia (Niez 2010) cannot be replicated with a standardised approach because public subsidies are often coupled with country-specific development objectives. However, following what is suggested by Bonan et al. (2017), some general features of the access expansion framework in which subsidies are supplied are relevant for all policymakers involved in such efforts:

- An adequate and effective implementing agency, able to operate autonomously from political pressure and being accountable through *a priori* established targets and measurable parameters must be in place.
- The electrification plan should be designed so as to be appropriate to the real needs and financial commitment possibilities of individuals, while also ensuring that other prior necessary conditions for its achievement are being met.
- A sound tariff policy which guarantees the financial sustainability and cost recovery of infrastructure development schemes, and thus that considers the

Table 5.7 Off-grid situation: rural electrification plans, standardised PPAs, and connection subsidies

Country/Source	Rural electrification plan	Standardised PPAs for off-grid power	Subsidies for grid connection
Burundi			
Kenya			
Malawi			
Mozambique			
Rwanda			
Tanzania			
Uganda			
South Africa			

Source: Climatescope (2017); various governmental websites
Red: Weak/absent policy framework; *yellow*: Relevant measures have been implemented; *green*: Functioning policy framework

actual measured ability-to-pay of households, must be established, rather than a politically agreed-upon figure.

Group-based subsidies, linked to number of applicants so to create a critical mass of customers and foster positive network and scale externalities, could also have a positive impact in contexts of relative poverty. On the other hand, electricity utilities should make the effort of unilaterally lowering their connection-related costs (and thus the derived consumer charges) by e.g. decreasing their material and management costs and adjusting technical standards to reflect the actual amount of electricity planned to be used by each household. The key idea is that public policy is a pivotal determinant of demand-side outcomes for energy access, while energy companies can also be flexible in changing their business models so as to unlock loss-loss traps.

Besides subsidies, energy access entails different additional policy dimensions, such as the need for a coordination plan (in particular in the context of rural areas), and the establishment of standardised PPAs to foster local private companies' investment in decentralised power infrastructure. Table 5.7 summarises the country-by-country situation for some of the key electrification policy frameworks.

To start with, a rural electrification plan is found virtually in every country.

- In Burundi the *Vision Burundi 2025 plan* has been approved in 2011, and the national objective is that of achieving an electrification rate of 25% by 2025. To accomplish this, Burundi has adopted a Decentralized Rural Electrification Strategy in 2015. However, the plan does not include specific objectives and clear supporting policy measures, but it rather represents a governmental vision.
- In Kenya, two plans have been set forth: the Rural Electrification Master Plan and the Distribution Master Plan. The former sets the objective of achieving a 65% access to electricity by 2022 and full access by 2030. Clear steps and supporting

instruments are defined. The Rural Electrification Authority is the main authority responsible for tracking progress of the plan. The latter, put forward in 2013, produced estimates of the long-term annual investment required in all distribution infrastructure, from 66 kV to LV, up to 2030.

- In Malawi, the Malawi Rural Electrification Program (MAREP) was last updated in 2017 and is currently undergoing its eight phase since its inception in 1980. It is supported by the Malawi Energy Regulatory Authority (MERA) and Rural Electrification Management Committee. Phase 8 foresees connecting to the grid 336 new trading centres by the end of 2018, along with generation capacity additions.
- In Mozambique at the end of 2017 a $500 million electrification program based on hydro, solar, micro-grids was launched by FUNAE. It aims at powering 332 villages through hydropower mini-grids with a combined capacity of about 1.01 GW and implementing 343 solar PV projects.
- Rwanda's Ministry of Infrastructure put forward in 2009 its Electricity Access Roll Out Program, which is being implemented by the national Rwanda Energy Group. This has been supported by 377 million USD in its first stage, which successfully increased electricity connection by 250,000 units in just 4 years. The second phase is currently ongoing, and it is being backed up by 300 million USD to achieve a 70% electrification rate.
- Tanzania's Power System Master Plan was last updated in 2016, has a planning horizon of up to 2040, and is being followed by TANESCO and EWURA. It suggests a 46 billion USD implementation cost and aims to boost power generation capacity to 10GW over the next decade. The government wants in fact to boost the electrification rate to 90% by 2035.
- Uganda approved its Indicative Rural Electrification Master Plan in 2009, which has been developed to reflect various alternatives of future network extensions, taking into account any planning for future transmission lines, sub-stations and distribution networks, industrial projects and international power exchange projects. The IREMP outlines guidelines, describes preferred standards and the phased implementation of future rural electrification in Uganda, as well as giving estimations of costs. It is intended that the IREMP act as a catalyst for the implementation of rural electrification projects. In 2015 a Grid Development Plan was also put forward to cover a period of 15 years and identify and justify new grid investments. It is reviewed and updated annually to reflect latest information on Government policy and strategies. It is also an input to the company's financial projections and annual budget.
- Finally, South Africa in 2013 approved its New Household Electrification Strategy (NHES), which replaced the Integrated National Electrification Programme (INEP). The NHES agreed upon defining universal access as 97% of households, as full electrification is unlikely to be possible due to growth and delays in the process of formalising informal settlements. 90% of households should be electrified through grid connection, while the rest with high-quality non-grid solar home systems or other possible technologies based on cost effective options in order to address current and future backlogs. The development of a master plan to

Table 5.8 RE feed-in-tariff situation

Country/Source	Wind	PV	CSP	Small hydro	Biomass	Geothermal
Burundi						
Kenya						
Malawi						
Mozambique						
Rwanda						
Tanzania						
Uganda						
South Africa						

Source: Climatescope (2017); various governmental websites
Red: FiTs absent; *green*: FiTs in place/drafted; *yellow*: FiTs under examination; *purple*: FiTs policy expired

increase efficiency in planning and the delivery process to ensure more connections is also expected. As of March 2018, the progress over the plan's objective stand at 84% for grid connections and at 66% for non-grid solutions.

As far as the situation of standardised PPAs for off-grid (in particular mini-grid) power generation is concerned, countries where standardised agreements for off-grid power purchase are offered include Tanzania, Uganda (where contracts also have sufficient duration and purchase obligation), while in Malawi, Rwanda and Kenya PPAs are in place but in a weaker form, and in Burundi and Mozambique they are completely missing from the national energy policy. Finally, while subsidies on the unit price of electricity for residential consumers are virtually in place in all countries, subsidies to lower grid connection barriers (e.g. loans for low-income families or direct subsidies for rural customers) are in place in Kenya, Rwanda, Tanzania, and Uganda.

Post-connection, feed-in-tariffs (Table 5.8) can be a further beneficial policy instrument in increasing private investment (from the scale of individual households and communities up to large companies), as they pay back the electricity-generating infrastructure owner who performed the upfront investment. However, if poorly set or managed, FiTs could also distort the market. In fact, they represent at the same time one of the most expensive way to subsidize RE and the single policy instrument that in many instances led to the quickest deployment. Thus, tariffs must be set in place appropriately, i.e. considering their burden on public finances, and adapted from time to time to keep up with changing socio-economic circumstances. On this point, Cox and Esterly (2016) remarked how conventionally FiTs for renewables have been set as fixed (e.g. per kWh produced), while in order to align with specific policy goals, policymakers should also consider varying FiTs payments by technology, project size, location and resource quality, as it is the case of Uganda's GET FiT program.

With regards to feed-in-tariffs:

- In Kenya FIT policy was adopted in March 2008 and revised first in January 2010 and then again in December 2012. The policy covers biogas, biomass, geothermal, small-hydro, solar and wind. Tariff differentiation is technology specific as well as size specific (below or above 10 MW). Hitherto, the policy has resulted in three IPP projects in Kenya, including a 29 MW biomass generator, a 40 MW geothermal project, and a small hydro site. On average, FiTs in Kenya payed 0.09-0.10 USD/kWh.
- Uganda's initial FIT policy was developed in 2007, under the Renewable Energy Policy on 2007. It was applicable up to 2009 and is referred to as REFIT phase 1. However, very limited uptake by developers was witnessed during the initial 3-year period. As a result, the policy was revised in 2012 and new tariffs were developed based on updated levelized costs of production so as to fast-track 20-25 small-scale renewable projects with a target capacity of 150 MW. The scheme, known as GET FiT, originally focused only on small hydro, bagasse (mostly sugarcane waste) and other biomass. In 2014, solar PV was included in the list of eligible technologies. To encourage participation, developers benefited from a top-up on the existing feed-in tariff and standardisation of PPAs and the development process. By the end of 2015, 17 GET FiT projects had been approved for 20 MW of biomass, 117 MW of small hydro and 20 MW of solar PV, with the first solar project commissioned in 2017, and others to follow in 2018. Currently GET FiT is under its third phase and it will offer support for up to 295 MW of hydro, bagasse and wind projects. Challenges remain to bring these projects online, as grid capacity is constrained and connections to the grid suffer from delays.
- In 2012, the Malawi Energy Regulatory Authority (MERA) drafted a feed-in tariff plan including small scale hydro (between 0.5 and 10 MW), PV, biomass, wind, and geothermal. For instance, for the case of hydro the tariffs apply for 20 years from the date of the first commissioning of the plant and they range between 0.08 and 0.14 USD/kWh depending on the project's scale and on the nature of the investor (firms or individuals), while for PV generation the FiT stands at 0.20 USD/kWh, for biomass and geothermal at 0.10 USD/kWh, and for wind at 0.13 USD/kWh. The policy also states that the FiTs shall be subject to review every 5 years from the date of publication. Any changes that may be made during such reviews shall only apply to RESE power plants that shall be developed after the revised guidelines are published.
- Decree 58/2014 created Mozambique's feed-in tariff, which applies to biomass, wind, small hydro and solar projects from 10 kW to 10 MW. Prices vary according to technology and capacity (ranging between 0.07 USD/kWh for large biomass projects up to 0.22 USD/kWh for solar PV up to 10 kW). According to this Decree, all projects must sell electricity to the state-owned utility EDM. Although the decree is available, injection of power into the grid cannot happen yet as some regulation is still to be approved. However, as of mid-2017, the FiT mechanism was already under revision. Mozambique is also

reviewing the scope of the National Electricity Council (CNELEC), the power market regulator, in order to broaden and strengthen its role. While this might help FiT mechanism to gain more space in Mozambique power market, however, the mechanism still has a long road ahead to be fully in force, since EDM, the only off-taker of all power contracts, is under considerable financial strain. Rwanda's Renewable Energy Feed-in tariff regulation was promulgated in February 2012.

- The Rwanda tariffs apply to small hydro from 50 kW to 10 MW. Contract terms are only 3 years and the law specifies that the tariffs cannot be reduced, while they are subject to a revision in the second year of the program to be implemented during the third year. Electricity tariffs are relatively high in Rwanda; however, a FIT policy is important to guarantee investors in renewable sources a ready market and an attractive return on investment for the electricity they produce. The FIT policy is still relatively new, but it has attracted the interest of RE developers.
- In Tanzania there are currently no FiTs in force. In 2014 a study has been commissioned by the Energy and Water Utilities Regulatory Authority (EWURA), as part of its collaboration with the US "Power Africa" scheme. However, as of late 2018 no regulation has been passed.
- To conclude, the case of South Africa's is of particular interest because it provides a relevant example to policymakers in EA countries of the kind of problems which might arise from the introduction of FiTs, as well as why shifting to a competitive tendering procurement process might be an effective alternative. South Africa approved RE-FiTs in 2009. They were designed not only to cover generation costs, but also to provide a real, after tax, fully indexed for inflation return on equity of 17%. Tariffs were initially particularly high, offering 0.156 USD/kWh for wind, 0.26 USD/kWh for solar PV, and 0.49 USD/kWh for solar CSP. After 2 years, the national regulator NERSA called for lower feed-in tariffs, arguing that a number of parameters, such as exchange rates and the cost of debt had changed. Tariffs were then adjusted downward, up to -41% for PV projects. At the same time, the Department of Energy and National Treasury concluded that the FiTs amounted to non-competitive procurement, and therefore were infringing public finance and procurement regulations. After several rounds of discussion and consultation, in 2011, the Department of Energy announced that the Renewable Energy Independent Power Procurement Program, a competitive bidding process for RE, would be launched, while NERSA officially abolished the RE-FiTs. During the 2 years of FiTs no PPA was signed, no procurement process was implemented, and the required contracts were never negotiated or signed. Competitive tenders introduced by the new regulation initially envisioned the procurement of 3.6 GW over up to five tender rounds, with a group of international and local experts assessing the bids. Over the four bidding rounds, USD 19 billion have been invested in 92 projects totalling 6.3 GW. As commented by Eberhard and Kåberger (2016), the competitive bidding process lead to increased competition (with record lows of 0.047 USD/kWh for wind and 0.064 USD/kWh for solar PV), and this was the main driver for steadily falling

prices over the bidding rounds. At the same time transaction costs kept falling in subsequent rounds, as the project sponsors and lenders became familiar with the REIPPPP tender specifications and process.

5.4.3 Payment Schemes and the Role of Digital Technologies

Finally, progressive tariffs,[3] lifeline tariffs,[4] microcredit and smart payment options via mobile phones all represent additional options with a strong potential to overcome the inability-to-pay. Governmental actions and plans for energy development should in fact also be drawing insights from research at the intersection between energy and behavioural economics (Spalding-Fecher et al. 2002; Pollitt and Shaorshadze 2011; Frederiks et al. 2015), so as to tackle some common issues such as split incentives; stalemate loss-loss situations; questions relating to energy efficiency; behavioural lock-ins; and thus nudge (Sunstein and Thaler 2008) individuals towards better energy decisions. It is important to remark that finding effective and secure payment and incentives solutions also has a feedback effect on the supply side because it fosters investments in capacity additions and grid infrastructure development from the private sector. Energy policy (in particular for mini-grid projects) should also be tightly linked with support schemes to local entrepreneurship: a coordination of the two could provide the chance for promoting productive energy uses even in remote rural areas, and thus pave the way out of energy-poverty traps.

In this sense, an emerging aspect deserving attention is the role that digital technologies are already playing in fostering energy development in the region. EA is indeed the first region in Africa for mobile-based services usage (WRI 2017). Mobile phones have rapidly spread as devices serving a multitude of functions beyond interpersonal communication, such as money transfer, bills payment, or access to bank services. Excluding South Africa, between 2005 and 2016 mobile phone subscriptions per 100 people in EA countries grew from an average of 5.7 to 60.7 (World Bank data), i.e. they witnessed an 11-fold multiplication (see Fig. 5.12). Considering the population growth pace and the large share of very young people in the EA, such trend seems even more dramatic.

In the last few years mobile phones have also gained increasing relevance for their potential breakthrough impact on different dimensions of energy access, including infrastructure planning and its operations, new business models, payments schemes, monitoring, data collection, and analysis. For instance, pay-as-you-go mechanisms are emerging as successful approaches to enable access in peri-urban

---

[3]Progressive tariffs result in a higher unit cost/kWh to customers who consume more electricity, so as to favour poorer household and encourage the fulfilment of most basic needs.

[4]Under lifeline tariffs, richer consumers cross-subsidise with their bills households that cannot afford to pay the market price of electricity.

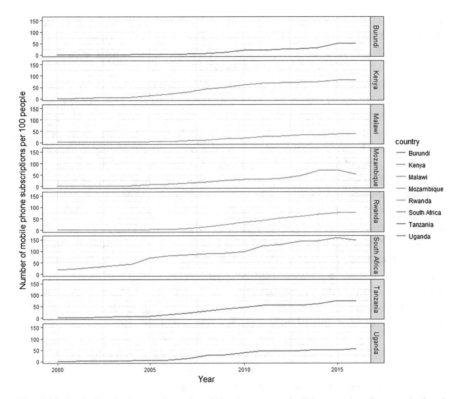

Fig. 5.12 Evolution in the number of mobile phone users in EA countries. Source: Authors' elaboration on World Bank data (2017)

and rural areas where investment was previously seen as both too costly by the demand-side and too risky by the supply-side. These systems allow households to rent the use of power generation infrastructure when they need energy, without having to purchase it *ex ante* with large upfront investments, or to purchase it in weekly or monthly instalments with variable leasing length while already benefitting from it. Payments become easy and immediate thanks to mobile phone networks (which in SSA are often found even in areas where the electricity grid is absent), and monitoring consumption levels and accepting payments hence turn into an inexpensive and transparent task for private providers. QR code-based application allowing to accept payments from customers' mobile phone without using point of sales machines, and thus further cutting on costs, are also being piloted. For example, Off Grid Electric's customers pay approximately $6 per month for entry-level systems, and $15–20 per month for small business kits that include various appliances.

Hitherto, much of the pay-as-you-go solar power innovations have focused on EA, and primarily Kenya. This is partly because of the popularity of mobile money in the region. Among the larger companies in the sector, there figure Off-Grid Electric, d.Light, Bboxx, Mobisol, Nova Lumos and M-Kopa Solar. Collectively, these have raised in excess of 360 million USD and they deployed more than 100,000 solar systems across the continent. They rely on IT systems—including web platforms, mobile apps and two-way SMS—to communicate with their customers and to manage power provisioning devices access and operations.

On the other hand, pay as-you-go entrepreneurs still represent an oligopoly, with just 4-5 companies active in the largest markets (Tanzania and Kenya), most of which are owned, managed and financed by foreign investor. In particular, WRI (2017) found 52 foreign private sector organizations backing such companies. This is partially due to the fact that local commercial banks are not willing to lend to local business because of the perceived risks of such new business models, and, as a result, local entrepreneurs struggle to access the capital they need to get started. In this sense, public finance from development finance institutions could unlock the situation and enable a larger competition field in the market, with a win-win for customers, electrification targets progress, and the local economy.

However, given the myriad of potential business plans and solutions that can be offered, companies should be addressed by supporting policy so as to maximise impact. These include payment schemes that foresee per-hour consumption costs, weekly flat tariffs with or without suspension and fines in case of discontinued payment. Randomized evaluation tests are undergoing (e.g. see Jack and Suri 2014) with the precise objective of finding out how price and payment methods affect the adoption of off-grid power-generation devices. Furthermore, a capillary data collection of consumption patterns allows public institutions to carry out of improved analysis to tailor additional investment and analyse expansion opportunities, paving the way for greater private-sector involvement in the market.

5.5   Channelling Investment: The Role of International Financing Institutions

The results of our analysis suggest that the average required capacity investment for matching demand growth beyond electrification ($80 billion)—is around the same of that needed for new electrification itself ($87 billion for a mean level between low, $61 billion, and high-tier, $113 billion, consumption).

Therefore, total investment cost for power generation, transmission, and distribution will amount to around $167 billion dollars between 2015 and 2030. Concerning the average investment required each year between 2016 and 2030, the figure corresponds roughly to $42.5/capita for the current EA population, or 2.6% of EA-7's GDP in weighted terms. International public finance institutions, such as multilateral development banks and national development agencies, could channel international private investments into Africa's power sector by putting in place dedicated blended finance tools and/or risk-sharing mechanisms.

In fact, the combination of political risks (e.g. corruption), commercial risks (e.g. solvability of consumers), lack of stable power market regulatory frameworks and lack of adequate power infrastructure, prevent international private investors from scaling-up investments in the continent.

International official development assistance (ODA) and other official flows (OOF) to the African power sector have tripled over the last decade, increasing from $2 billion in 2005 to $8 billion in 2015. The World Bank Group (WBG), the European Union (EU) (i.e. EU institutions + EU Member States) and the African Development Bank (AfDB) disbursed most of the funds in the sector, while actors like the United States (US), the Climate Investment Funds (CIF), the Arab Fund for Economic and Social Development (AFESD), the OPEC Fund for International Development (OPEC-FID) and others played a far minor role.

About 90% of the last decade's international financial support to Africa's electrification came from only three players: WBG, AfDB, EU. In terms of sectorial destination, the WBG mainly invested in non-renewable power generation (particularly coal), while the EU mainly invested in renewable power generation (namely hydro, wind and solar). On its side, the AfDB mainly invested in power transmission and distribution infrastructure.

It is also worthwhile to outline the geographical distribution of the various players' investments. The WBG and the EU are key players in sub-Saharan Africa (excluding South Africa), while the AfDB is the key player in South Africa (Fig. 5.13).

On their side, Chinese companies (90% of which state-owned) have also heavily invested into Africa's power sector. About 30% of new power capacity additions in sub-Saharan Africa between 2010 and 2015 was indeed financed by China, for a total investment about of around $13 billion over the quinquennium. Chinese contractors have built or are contracted to build 17 GW of power generation capacity in sub-Saharan Africa from 2010 to 2020, equivalent to 10% of the existing installed capacity in the region. In geographical terms, these projects are widespread across the region, and concerns at least 37 countries out of 54. In terms of capacity size, Chinese contractors primarily focus on large projects. In terms of type of capacity, they primarily focus on traditional forms of energy like hydropower (49% of projects 2010–2020), coal (20%) and gas (19%), while involvement in modern renewables remains marginal (7%).

Over the last years, to maximise impact and leverage on private investors, the WBG and the AfDB have streamlined their actions in the field, focusing resources on a few initiatives. The WBG operates through its established mechanisms (i.e., The International Bank for Reconstruction and Development, The International Development Association, The International Finance Corporation, The Multilateral Investment Guarantee Agency). The AfDB, in addition to its traditional financing tools, has established two initiatives to invest in Africa's (electricity) infrastructure: the 'New Deal on Energy for Africa' and the 'Africa50'. The former is a public-private partnership between the AfDB, African governments and global private sector aimed at putting in place innovative financing to achieve universal access to energy by 2025, while the latter is an infrastructure fund owned by the AfDB, African governments and global institutional investors created to specifically mobilize long term savings to promote (electricity) infrastructure development in Africa.

Fig. 5.13 Cumulative
financial assistance to Africa
power sector, by region
(2005–2015). Source:
Authors' elaboration on
OECD Development
Finance Database

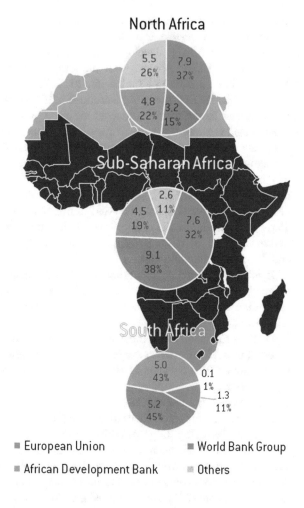

On the contrary, the EU's action appears to be particularly fragmented. The EU
has 26 initiatives ongoing in the field, originating from either EU Member States or
EU Institutions. The variety of EU Member States' initiatives is understandable, as
each country has its own political and commercial interests to promote across Africa.
What is less understandable is the fragmentation of EU Institutions. The EU's
current fragmented system seems to favour overlaps, inefficiencies and overall
higher transaction costs. It is reasonable to consider that European taxpayers'
money would be far better spent if channelled through a unique facility, allowing
policy consistency, elimination of overlaps, abatement of transaction costs and,
therefore, overall higher efficiency and impact. This could be done by coordinating
current and prospective EU programs though the recently-established 'EU External
Investment Fund'.

References

Abbey C, Joos G (2009) A stochastic optimization approach to rating of energy storage systems in wind-diesel isolated grids. IEEE Trans Power Syst 24:418–426. https://doi.org/10.1109/TPWRS.2008.2004840

Ackah I, Opoku FA, Suleman S (2017) To toss a coin or shake a hand: an overview of renewable energy interventions and procurement in selected African countries. MPRA paper 77489. University Library of Munich, Germany

AllAfrica (2017) Mozambique: electricity prices rise

Bernard T, Torero M (2015) Social interaction effects and connection to electricity: experimental evidence from rural Ethiopia. Econ Dev Cult Chang 63:459–484. https://doi.org/10.1086/679746

Bhatia M, Angelou N (2015) Beyond connections: energy access redefined. World Bank, Washington, DC

Bonan J, Pareglio S, Tavoni M (2017) Access to modern energy: a review of barriers, drivers and impacts. Environ Dev Econ 22:491–516. https://doi.org/10.1017/S1355770X17000201

Castronuovo ED, Usaola J, Bessa R et al (2014) An integrated approach for optimal coordination of wind power and hydro pumping storage. Wind Energy 17:829–852. https://doi.org/10.1002/we.1600

Conway D, van Garderen EA, Deryng D et al (2015) Climate and southern Africa's water–energy–food nexus. Nat Clim Chang 5:837–846. https://doi.org/10.1038/nclimate2735

Cox S, Esterly S (2016) Feed-in tariffs: good practices and design considerations

Creutzig F, Nemet G, Luderer G et al (2017) The underestimated potential of solar energy to mitigate climate change. Nat Energy 2:17140. https://doi.org/10.1038/nenergy.2017.140

Deichmann U, Meisner C, Murray S, Wheeler D (2011) The economics of renewable energy expansion in rural Sub-Saharan Africa. Energy Policy 39:215–227

Dinkelman T (2011) The effects of rural electrification on employment: new evidence from South Africa. Am Econ Rev 101:3078–3108

Eberhard A, Kåberger T (2016) Renewable energy auctions in South Africa outshine feed-in tariffs. Energy Sci Eng 4:190–193. https://doi.org/10.1002/ese3.118

Eberhard A, Gratwick K, Morella E, Antmann P (2017) Independent power projects in Sub-Saharan Africa: investment trends and policy lessons. Energy Policy 108:390–424

ESI Africa (2016) Exclusive interview with Eng. Mr. Lebbi Mwendavanu Kisitu Changullah, Secretary General of EAPP. https://www.esi-africa.com/features/exclusive-interview-eng-mr-lebbi-mwendavanu-kisitu-changullahsecretary-general-east-african-power-pool/

Expogroup (2017) East Africa: lighting East Africa. Expogroup. http://expogr.com/rwanda/lightexpo/detail_news.php?newsid=5019&pageid=2

Fitzgerald N, Lacal Arántegui R, McKeogh E, Leahy P (2012) A GIS-based model to calculate the potential for transforming conventional hydropower schemes and non-hydro reservoirs to pumped hydropower schemes. Energy 41:483–490. https://doi.org/10.1016/j.energy.2012.02.044

Fouquet R (2016) Path dependency in energy systems and economic development. Nat Energy 1:16098

Frederiks ER, Stenner K, Hobman EV (2015) Household energy use: applying behavioural economics to understand consumer decision-making and behaviour. Renew Sust Energ Rev 41:1385–1394. https://doi.org/10.1016/j.rser.2014.09.026

Golumbeanu R, Barnes D (2013) Connection charges and electricity access in Sub-Saharan Africa. https://doi.org/10.1596/1813-9450-6511

Grey D, Sadoff C, Connors G (2016) Effective cooperation on transboundary waters

Grimm M, Lenz L, Peters J, Sievert M (2017) Demand for off-grid solar electricity—experimental evidence from Rwanda

Höök M, Aleklett K (2010) A review on coal-to-liquid fuels and its coal consumption. Int J Energy Res 34:848–864. https://doi.org/10.1002/er.1596

IEA (2011) World energy outlook 2011

IEA (2017a) World energy outlook 2017

IEA (2017b) WEO 2017 special report: energy access outlook. International Energy Agency

Ikejemba ECX, Mpuan PB, Schuur PC, Van Hillegersberg J (2017) The empirical reality and sustainable management failures of renewable energy projects in Sub-Saharan Africa (part 1 of 2). Renew Energy 102:234–240. https://doi.org/10.1016/j.renene.2016.10.037

IndexMundi (2018) Coal, South African export price - monthly price. https://www.indexmundi.com/commodities/?commodity=coal-south-african&months=60

IPCC (2015) Fifth assessment report—the physical science basis.. http://www.ipcc.ch/report/ar5/wg1/. Accessed 17 Jan 2018

IRENA (2012) Renewable energy technologies: cost analysis series, hydropower

IRENA (2016) Solar PV in Africa: costs and markets. http://www.irena.org/publications/2016/Sep/Solar-PV-in-Africa-Costs-and-Markets. Accessed 23 Apr 2018

IRENA (2017) Electricity storage and renewables: costs and markets to 2030. http://www.irena.org/publications/2017/Oct/Electricity-storage-and-renewables-costs-and-markets. Accessed 6 Aug 2018

IRENA (2018) Renewable power generation costs in 2017 /publications/2018/Jan/Renewable-power-generation-costs-in-2017. /publications/2018/Jan/Renewable-power-generation-costs-in-2017.. Accessed 3 Aug 2018

IWACU Burundi (2017) Power price to increase in Burundi

Jack W, Suri T (2014) Risk sharing and transactions costs: evidence from Kenya's mobile money revolution. Am Econ Rev 104:183–223. https://doi.org/10.1257/aer.104.1.183

Kammen DM, Jacome V, Avila N (2015) A clean energy vision for East Africa: planning for sustainability, reducing climate risks and increasing energy access. https://rael.berkeley.edu/wp-content/uploads/2015/03/Kammen-et-al-A-Clean-Energy-Vision-for-the-EAPP.pdf. Accessed 23 Apr 2018

Khandker SR, Barnes DF, Samad HA (2013) Welfare impacts of rural electrification: a panel data analysis from Vietnam. Econ Dev Cult Chang 61:659–692

Lee K, Brewer E, Christiano C et al (2016) Electrification for "Under Grid" households in Rural Kenya. Develop Eng 1:26–35. https://doi.org/10.1016/j.deveng.2015.12.001

Mas'ud AA, Wirba AV, Muhammad-Sukki F et al (2016) A review on the recent progress made on solar photovoltaic in selected countries of sub-Saharan Africa. Renew Sust Energ Rev 62:441–452

Mekmuangthong C, Premrudeepreechacharn S (2015) Optimization for operation and design of solar-hydro pump energy storage using GAMS. Appl Mech Mater 781:346–350. https://doi.org/10.4028/www.scientific.net/AMM.781.346

Mentis D, Howells M, Rogner H et al (2017) Lighting the world: the first application of an open source, spatial electrification tool (OnSSET) on Sub-Saharan Africa. Environ Res Lett 12:085003

Millenium Challenge Corporation (2017) Malawi's power sector reforms spur private-sector participation and utility turnaround

Moreno A, Bareisaite A (2015) Scaling up access to electricity: pay-as-you-go plans in off-grid energy services

Murage MW, Anderson CL (2014) Contribution of pumped hydro storage to integration of wind power in Kenya: An optimal control approach. Renew Energy 63:698–707. https://doi.org/10.1016/j.renene.2013.10.026

Namara RE, Giordano M (2017) Economic rationale for cooperation on international waters in Africa: a review. World Bank

Niez A (2010) Comparative study on rural electrification policies in emerging economies

Nyasa Times (2018) Malawi electricity tariff up by 24.6%

Occhiali G (2016) Power outages, hydropower and economic activity in Sub-Saharan Africa. PhD thesis, University of Birmingham

Parshall L, Pillai D, Mohan S et al (2009) National electricity planning in settings with low pre-existing grid coverage: development of a spatial model and case study of Kenya. Energy Policy 37:2395–2410. https://doi.org/10.1016/j.enpol.2009.01.021

PBL Netherlands Environmental Assessment Agency (2017) Towards universal electricity access in Sub-Saharan Africa. PBL Netherlands Environmental Assessment Agency. http://www.pbl.nl/en/publications/towards-universal-electricity-access-in-sub-saharan-africa. Accessed 23 Apr 2018

Pillot B, Muselli M, Poggi P, Dias JB (2017) On the impact of the global energy policy framework on the development and sustainability of renewable power systems in Sub-Saharan Africa: the case of solar PV. arXiv preprint arXiv:170401480

Pollitt MG, Shaorshadze I (2011) The role of behavioural economics in energy and climate policy. Faculty of Economics

Renewable Energy World (2017) What is the business case for energy storage in Africa?. https://www.renewableenergyworld.com/articles/2017/11/what-is-the-business-case-for-energy-storage-in-africa.html. Accessed 3 Sep 2018

RISE (2017) RISE renewable indicators for sustainable energy

Sloss L (2017) Emissions from coal-fired utilities in South Africa and neighbouring countries and potential for reduction

Spalding-Fecher R, Clark A, Davis M, Simmonds G (2002) The economics of energy efficiency for the poor—a South African case study. Energy 27:1099–1117. https://doi.org/10.1016/S0360-5442(02)00081-6

Stima Consultancy (2018) Electricity cost in Kenya

Sunstein C, Thaler R (2008) Nudge. The politics of libertarian paternalism. Yale University Press, New Haven

TANESCO (2016) TANESCO tariff adjustment: application for year 2017

Techjaja (2018) These are the latest UMEME and Yaka Power tariffs 2018

The East African (2018) Boost for EA power pool project as Dar, Nairobi secure $600m. The East African. http://www.theeastafrican.co.ke/business/Boost-for-East-Africa-power-pool-project/2560-4642124-tgp4b6/index.html

The New Times (2017) Officials explain recent power pricing

The World Bank (2017) World Bank Data.. Accessed 20 Nov 2017

Transboundary Freshwater Dispute Database (2017) Transboundary freshwater spatial database. In: Program in water conflict management and transformation. Oregon State University, Corvallis. https://transboundarywaters.science.oregonstate.edu/content/transboundary-freshwater-spatial-database

Turner SWD, Hejazi M, Kim SH et al (2017) Climate impacts on hydropower and consequences for global electricity supply investment needs. Energy 141:2081–2090

UNIDO (2017) Renewable energy-based mini-grids: the UNIDO experience|ECOWREX. http://www.ecowrex.org/document/renewable-energy-based-mini-grids-unido-experience. Accessed 6 Aug 2018

World Energy Council (2017) Energy resources. https://www.worldenergy.org/data/resources/resource/coal/

WRI (2017) "Pay-As-You-Go" solar could electrify rural Africa

Chapter 6
RE Interaction with NG Resources

Contents

A highly relevant question in the discussion on the energy development of EA concerns the relationship that will materialise between RE and the abundant NG resources that are undergoing production in the region. Already in previous chapters it was evidenced that a configuration of RE-NG complementarity could become the least-cost and most sustainable pathway: (1) to guarantee a full electrification of EA; (2) to match the steeply growing demand for power to feed an emerging industrial sector; and (3) to contribute to the objective of clean cooking. In particular, gas could indeed contribute to the displacement of coal in electricity generation, and thus allow for substantial potential greenhouse gas emissions' reductions.

Reserves are concentrated along the coasts of Mozambique and Tanzania, and they are estimated in the two countries at 2832 bcm and 1614 bcm, respectively. Extraction has begun in the 2000s, and it has reached a total of 6.80 bcm in year 2015 (Fig. 6.1), with the bulk of it produced in Mozambique.

At the same time, development of LNG liquefaction units is undergoing in both countries. As pointed out by Demierre et al. (2015), in Mozambique this move has been a consequence of the government strategy of establishing contracts with international oil and gas majors with the chief aim of achieving a net exporter positions, especially towards the Asian market. As seen in Fig. 6.2, ever since the inception of NG production, Mozambique has been exporting most of it. Conversely, hitherto Tanzania focused on developing gas-to-power plants (with 862 MW of gas-fired capacity added since 2004) to increase domestic generation and diversify its power mix away from hydropower.

In this context, a small profit margin for exports due to low international LNG market prices (standing around 7 \$/MBTU for European prices (UK NBP) and around 10 \$/MBTU for Far East prices (JKM) as of February 2018), as well as increasing domestic energy demand in EA, render a discussion of the role that NG could play within the boundaries of the region particularly relevant. Here, with reference to the least-cost electrification analysis of Chap. 4 and to the technological

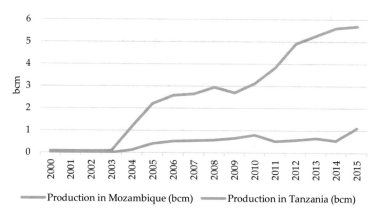

Fig. 6.1 Evolution of NG production in EA. Source: Author's elaboration on US EIA data

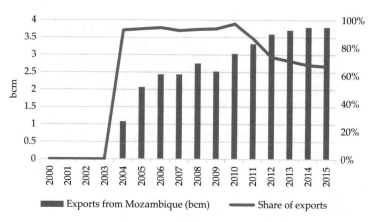

Fig. 6.2 Evolution of NG export in Mozambique (absolute amount and as share of total production). Source: Author's elaboration on US EIA data

questions constraining in different ways RE, we focus on the question of the potential complementarity that gas resources could exhibit with the domestic development of RE.

1. The use of NG is still limited in the region as a share of the total primary energy supply (IEA 2017), although gas-to-power gained a relevant share in the generation mix of Tanzania. However, a scenario where in the coming years NG will reach a significant degree of penetration in the regional energy market is deemed possible. This has important implications for RE potential development, as it could either complement it (e.g. providing peak power generation potential to support hydro and solar), or take up a large share of total installed baseload capacity, and hence limit the large-scale deployment of RE sources as the main suppliers of baseload power and set the region to a higher emissions trajectory.

2. A similar concern can also be relevant for an alternative scenario of rising international market prices for NG and LNG, with the disregard of the resource for domestic use and its consequent export. This could have side-effects, such as the employment of gas export revenues for the development of coal-fired baseload power plants and development of domestic and import of foreign coal resources.

3. A third consideration is that excessive (i.e. economically inefficient) rates of RE penetration could be witnessed, for instance as a result of a strong public policy-push. This could result in the displacement of NG and in a shift of the resource on the market for international exports away from EA, even in settings where gas-fired generation would be the least-cost and most cost-effective option.

Overall, these issues entail discussing the question of how to embark the regional energy mix on an optimal pathway through infrastructure investment and planning. On the one hand, gas extraction processes and the technology for its distribution and conversion to electricity are well established and highly adaptable to different needs and scales. The greatest benefit of NG is that it does not *per se* present intermittency issues. Also, the electricity output of gas-fired power plants is relatively easier to ramp-up or down than that for coal-fired steam power plants. This allows prompter and more efficient adjustments of the supply to demand fluctuations and therefore a better complementarity with intermittent renewable sources.

Despite being abundant, RE potential (and mostly solar PV and wind power) is in fact limited by a number of technical and economic questions, with the pivotal being intermittency. This depends both on natural cycles such as the day-night alternation or the seasons, and on scarcely predictable climate weather events such as wind speed (the variability of which will also increase with climate change). Furthermore, as discussed in the previous chapter, notwithstanding declining costs, battery technologies for power storage are not yet competitive enough to ensure a continuous coverage of electricity needs through RE only. The issue is further exacerbated by the fact that the electricity demand from different sectors, and in particular in developing economies, can be rather unpredictable at different time scales. Looking at other RE sources, geothermal and hydropower do not present harsh intermittency issues. However, geothermal is not everywhere sufficiently available to satisfy the growing demand for power in EA countries, and even when this is the case (e.g. Kenya), it requires costly explorations to identify sites with potential. Hydropower faces instead issues of a different nature: generation is dependent on stream flows and reservoirs levels, which in turn depend on rain intensity and frequency, as well as on evapotranspiration and on water withdrawals for other uses, such as agriculture, domestic consumption, and industrial uses, while it also presents an international cooperation dimension.

Gas is also poly-functional: it can be employed for both electricity and heat generation, or in industrial processes, or in fostering the transport sector in urban areas, as well as to support fertilizer production for agriculture. Overall, the greatest contribution of NG could be in urban areas, which are quickly growing in size and importance in EA (United Nations Population Division 2017) and where the population density is high, and the industry and transport sector require increasingly large energy inputs. This includes the development of a number of gas-fired power plants to support the large-scale deployment of renewables, as well as of an interconnected high-voltage international grid system to compensate temporary country-level imbalances. Both these measures certainly represent positive contributions to the achievement of electricity access security in EA.

The coupling of access to electricity objectives with the establishment of a distribution network of energy sources for clean cooking and for the local industry, e.g. LPG tanks for cooking or NG pipelines, deserves attention. Many examples are emerging for the former case, with the Kenyan company KOKO Networks using technology to lower off-pipe fuel distribution costs by launching a network of cloud-connected stations in local shops, where consumers are able to buy modern cooking stoves, with which ethanol fuel is then purchased through a digital billing system in small bundles. In Tanzania, the local KopaGas has also launched a pay-per-use LPG model, allowing customers to use mobile money to pay-off gas and cooking appliances over time, at a cost which is competitive with charcoal market prices. Furthermore, in the proximity of urban areas where gas pipelines are either under construction or planned, pre-paid gas supply and smart metering approaches are gaining relevance, with companies exploring distribution options.

A number of issues put nonetheless an upper bound to the likelihood of a scenario of region-wide domestic deployment of NG, and thus go against the hypothesis that gas could extensively displace renewables for baseload generation. First, distribution infrastructure (pipelines) is very capital intensive: there exist distribution constraints to domestic consumption in EA-8, and in particular in rural regions where the distance from the grid is long and the population is scattered in villages of low density. In these rural areas, where currently the bulk of the population is concentrated, renewables already have a strong comparative advantage and are unlikely to be displaced in the future. Financing is undoubtedly an issue: Demierre et al. (2015) estimated the total capital cost of the gas transmission infrastructure in the EA region at approximately \$57 billion in their baseline scenario. The collection of such amounts of capital depends largely on the contribution of international development banks (which however today appear reluctant to finance large-scale fossil fuel projects), as well as of private foreign direct investment (FDI). In turn, the latter is related to multiple institutional challenges, not last the establishment of a coordinated plan of regional cooperation and infrastructure sharing. Furthermore, while gas-fired generation is not intermittent *per se*, it does pose back up concerns for reliability: capacity factors of NG combined cycle plants stand at an average of 50% (US EIA 2017), and thus NG generation requires redundancy to secure a constant supply.

Upon these considerations, we argue that under the current and future forecasted conditions, a diversification of the energy mix is likely to render the power sector of EA more resilient to economic and other exogenous factors, and thus have a positive role in strengthening energy security in EA. Solar and wind power have large technical potential, as well as a guaranteed sustainability in the long-term horizon and large financial support of international development institutions; hydropower is abundant in different countries, but it is likely to be affected by changes in climate patterns (and by intensifying extremes such as flooding and drought events), and it requires temporary water resource withdrawal, potentially determining multi-sector competition. Under a sustainability-oriented and long-sighted governance—gas resources represent an effective match to renewables, as gas could provide a solid back-up to variable REs. Furthermore, an emerging industrial sector needs abundant and secure supply of power and heat to guarantee that the development process takes off. In this sense, hybrid gas-renewable systems like integrated solar combined cycle allow to tap into the increasing combined demand of electricity and heat from industry.

Overall, it is important that NG resources in the region support energy development at a low-carbon trajectory. We believe that they have the potential to displace imports of coal and diesel/HFO back-up generation, and thereby help the region to avoid taking a high-carbon energy development pathway and lock-in. Conversely, the domestic development of NG resources should be planned in tight complementarity with RE, given the abundance and the learning that is taking place in the latter sector. Together, RE and NG can represent the bulk of the new capacity additions in the coming years (IEA 2017), and together they can play an important role in complementing the shortcomings of each other, including intermittency and uncertainty in output, while also feeding the productive sectors of the economy and powering urban areas.

References

Demierre J, Bazilian M, Carbajal J et al (2015) Potential for regional use of East Africa's natural gas. Appl Energy 143:414–436. https://doi.org/10.1016/j.apenergy.2015.01.012
IEA (2017) World energy outlook 2017
United Nations Population Division (2017) World population prospects: the 2017 revision
US EIA (2017) International energy statistics. https://www.eia.gov/beta/international/data/browser/#/?c=4100000002000060000000000000g00020000000000000001&vs=INTL.44-1-AFRCQBTU.A&vo=0&v=H&end=2015. Accessed 23 Apr 2018

Chapter 7
Conclusions and Policy Implications

This book has investigated the potential of RE to empower an energy development process in EA, the key challenges to its exploitation, and its relationship with other energy sources. We have touched upon the reasons behind the lagging behind of EA-7 countries in energy development terms, and we have discussed how their technical potential could be effectively turned into installed capacity. At the same time, we have also highlighted a number of technical, economic, cooperation, policy, and financing challenges which must be tackled to achieve such objective.

Chapters 2 and 3 highlighted that EA-8 is a region with abundant energy resources, and that these would be technically sufficient to guarantee a self-sufficient development of the regional energy sector, and thus of the overall regional economy. Solar (and in particular PV potential) is the most evenly widespread option and in many areas and it also represents the cheapest solution. Additional generation potential—both as RE (mostly hydropower and geothermal) and fossil (NG and coal)—is abundant but heterogeneously scattered in the eight countries considered. However, irrespective of such large potential, all EA-7 countries are still lagging behind in terms of their energy and economic development potential. The EA-8 regional installed capacity stands at 55 GW, which become only 8 GW when South Africa is disregarded. Among these, most capacity is found in Kenya, Tanzania and Mozambique, although the bulk of power generated in the latter is exported to South Africa. This is rather striking for an overall population of 271 million people (215 without South Africa).

Chapter 4 elaborated on the key challenges of least-cost electrification in EA-7, and it found that achieving universal electricity access targets set by the Sustainable Development Goal 7 and satisfying the growth in the demand from already electrified consumers and other sectors would require a mean total investment of about $167 billion between today and 2030. Results also showed that while a coal-based expansion scenario will have slightly lower upfront investment requirements than one based on RE-NG, it will also become the costliest over the long-run. Conversely, RE-NG development will guarantee substantially lower future costs and greenhouse gas emissions. With regards to the different investment components, it resulted that bringing access to electricity to the entire population of the region would require investments in the $60–113 billion range (corresponding to a median of $87 billion, or $5.8 billion/year until 2030) depending on the consumption tier considered and on different projections over shifting cost profiles. When the projections over the growth in the demand for power of already electrified households and an emerging industry sector are also accounted for, a further mean investment of about $80 billion (split roughly equally between power generation and grid infrastructure expansion) is calculated as the requirement to satisfy the demand-side. Putting the mean total of $167 billion in perspective shows that it corresponds to an annual investment of 2.6% of weighted GDP—defined as the weighted average of PPP GDP by the share of international-targeted energy access investment and of exchange rate GDP by the share of local-targeted investment – or to an annual average of $42.5 per-capita during the 15-year period under examination. The mean annual required investment for new electrification stands instead at $37 per person currently without access (also accounting for population growth dynamics), or an annual 1.3% share of today's weighted GDP. Such disaggregated results help developing a better understanding of the implications and required action from households, private companies, and international organisations. For instance, they reveal that—since 34% of the population of EA-7 is living below the poverty line of around $2/day/person ($730 per year)—a large part of the investment gap stems from the inability to pay of the population currently without access. Cross-subsidies targeted at closing the energy access gap and developing the necessary infrastructure seem therefore a key measure to be implemented from local policymakers. In turn, over the long-run the achievement of universal access to electricity would stimulate the demand in all sectors and boost economic growth of countries, with widespread returns also to those who provided the capital to perform infrastructure investments.

Given such conclusions, Chap. 5 then discussed why governmental action is highly needed to create a suitable investment environment to channel such capital, in particular for the domestic private sector and international institutions. Policies in support of RE are only a part of a wider market design that requires rules to govern the interaction of all the economic agents involved in power generation, transmission, distribution and retailing. It was outlined how risks arising from macroeconomic (e.g. exchange and interest rates, or inflation) or political instability, or from weak protection of contract and property rights, as well as from the lack of a stable regulatory framework and due to underdeveloped financial institutions and markets (able to guarantee for the creditworthiness of the off-takers and ensure long and short-term financing availability) represent determinant aspects for energy development to take place.

To conclude, Chap. 6 discussed the challenges and opportunities deriving from the recent discovery and development of NG reserves along the coasts of Mozambique and Tanzania. The chapter highlighted challenges and opportunities in the realisation of a regional generation mix based on a RE-NG tandem, and it discussed the factors which might result in pushing the domestic development of each of the two sources. The role of the international LNG price was evidenced as a key decision variable for either the focus on exports or the regional gas market development, and the responsibility of governmental policy was highlighted in achieving an optimal and sustainable power mix.

On the basis of the book's analysis, the following policy implications can be formulated:

- If large-scale grid-connected RE projects are to be implemented, private investment is necessary, and IPPs are pivotal players. These should be incentivised to enter the market with competitive tendering processes following principles of efficiency and cost-effectiveness, rather than having a direct negotiation with government actors.
- All economically viable, locally available energy sources should be considered, not only for export, but chiefly for domestic consumption and regional distribution (accounting for their external costs, e.g. environmental damage, health-related issues, and climate impacts). Given the particular setting of EA, the focus should be on realizing the potential synergies between RE and NG, with the objective of achieving a least-cost and low-carbon electrification scenario. In all likelihood, RE will cover the greatest fraction of the demand if electrification is to be diffused. NG can serve major urban centres and their industry, foster the development of the transport and manufacturing sectors, and serve non-electric needs in rural areas (e.g. with LPG distribution). This entails the design of energy resources optimisation plans looking into the future, which account for economic, technical, and environmental aspects.
- The role of the Eastern Africa Power Pool (EAPP) for the energy development of the region should be boosted, and master plans long-sighted: the most effective strategies in terms of infrastructure development, market and contract design, international policy, import/export dynamics, and potential cooperation spill-overs should be identified and rendered operational. Innovative technological solutions such as storage, demand response and micro-grids could play a significant role in the process. Fostering regional interconnections could in fact allow to maximize national energy potentials and integrate more generation capacity into the system, while ensuring its robustness and an efficient management of RE intermittency. More interconnection would also imply the possibility to set the power sector of the region on a significantly lower emissions trajectory than if countries developed their power sector independently. Furthermore, cooperation in transboundary water basin management is of high significance in the context of increasing water scarcity and competition among sectors, in particular considering the issue of hydropower dependency in several EA countries.
- Political, regulatory, and security risks all represent barriers in achieving private project financing. Enabling institutional and market conditions must therefore be satisfied. This implies, among other things, that access to markets should be improved via roads, telecommunications and institutions so as to increase the economic impact and thus the profitability of electrification investments. The diffusion of digital technologies can play a major role in this sense, as these enable better infrastructure planning and operations, new business models capable of reaching larger shares of the population, easier and smarter payments

schemes, cheap monitoring, and big data collection and analysis for informing policy on future planning.

- According to our electrification scenarios, in the context of EA grid capacity additions and its extension will constitute the largest investment component, with a median value of 58% of total required investments. Nonetheless, those for mini-grid technologies (including PV, wind and hydro-based solutions) represent roughly one-third of total median investments, while those for standalone PV or diesel solutions account for less than 10%. Thus, the grid/mini-grid trade-off should be carefully evaluated in different regions while accounting for an array of conditions, impacts, and uncertainties. Clear rules over potential future interconnections should be set before mini-grid investments are performed, so as to reduce the uncertainty over the long-run prospects of such investment, and thus encourage private participation. Resilience of hard infrastructure over different future scenarios is key.

- Overall, solar PV is the RE with the largest penetration among mini-grid and standalone solutions, owing to its low cost and massive availability throughout the entire EA-7 region. On the other hand, wind and hydro mini-grids are competitive only in certain circumscribed areas with high potential. Diesel generators are in most cases not cost-competitive with RE to achieve the desired consumption tier targets.

- Subsidies schemes for connection to the grid or off-grid renewable energy infrastructure installation should be carefully designed and put in place to complement the low ability-to-pay of households and promote their socio-economic development. Communities largely relying on biomass or small diesel gen-sets for their energy needs may be unwilling to invest large proportion of their incomes to achieve RE-based electrification without a predictable return. In particular, the upfront costs for decentralised systems may still be higher than most consumers are willing or able to pay. Subsidies should however be set in the context of well-designed and comprehensive electrification and development plans which look at the specificities of the country, and not be driven by mere political forces. Subsidies are in fact burdensome for public finances, and thus should be spent optimally when in place. The establishment and enhancement of the role of independent energy authorities could be a further positive step in this direction.

- Renewable energy feed-in-tariffs can incentivise both individuals and companies to invest in infrastructure and have a guaranteed return into the future. In the case of IPPs procuring large power plants, FiTs provide price certainty and long-term contracts that stimulate and help finance renewable energy investments. Being a costly policy instrument, FiTs should be designed so as not to distort the market and to be flexible according to different technologies and changing conditions, and be complemented by tendering processes.

- For the effective development of off-grid technology to electrify rural areas, the implementation of a set of institutional conditions is highly relevant. Renewable Energy Authorities should be established where absent and have the sufficient discretion and autonomy to regularly update national electrification plans, lobby national governments to provide the means to implement such plans, make sure that private actors are encouraged to invest by offering long-term standardised PPAs and transparent rules (including for future interconnections), and offering connection grants, subsidies and incentives to potential customers.

- Electricity utilities' billing schemes should aim at having large flexibility to accommodate as much as possible the needs of credit-constrained households and to nudge them to gain access, so as to trigger a win-win scenario for the provider and for the consumer. Behavioural considerations (e.g. the way information is presented) and the default contract type can play a role in this sense, especially in the emerging context of mobile-based solutions and smart payment schemes.

- Small energy enterprises should cooperate with state-owned electricity companies to create integrated business-to-business arrangements and so increase access to potential markets and funding. The scale of decentralised energy systems often requires a special and context-specific focus on projects, business models and financial solutions which either of the two institutions could not be able to provide by itself.

Appendix A: Additional Tables and Figures

Burundi

Installed capacity and share of electricity generated, 2014 (AfDB)

Burundi	Installed capacity (MW)	Generation (GWh)
Total	41	304
Hydro	32 (78%)	253.3 (83.2%)
Thermal	9 (22%)	51 (16.8%)

Electricity consumption by sector, 2014 (UNdata)

Burundi	Sector's consumption (GWh)
Total	264.9
Manufacturing, construction and non-fuel industry	108.6 (41%)
Households	129.8 (49%)
Other	26.5 (10%)

M. Hafner et al., *Renewables for Energy Access and Sustainable Development in East Africa*, SpringerBriefs in Energy, https://doi.org/10.1007/978-3-030-11735-1

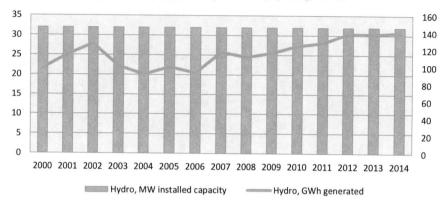

Hydropower installed capacity and generation, 2000–2014 (UNdata)

Kenya

Installed capacity and share of electricity generated, 2014 (UNdata)

Kenya	Installed capacity (MW)	Generation (GWh)
Total	2094.4	9138.6
Hydro	797 (38%)	3569 (39%)
Geothermal	558 (26.6%)	2917.4 (31.9%)
Thermal	734.1 (35%)	2635.2 (39%)
Wind	5.3 (0.3%)	17 (0.2%)

Installed capacity and share of electricity generated, 2015 (Energy Regulatory Commission)

Owner	Fuel	Capacity (MW)	Generation (GWh)
KenGen	Hydro	820 (36.1%)	3308 (36.2%)
	Thermal	263 (11.6%)	492 (5.4%)
	Geothermal	488 (21.5%)	3104 (34%)
	Wind	25.5 (1.1%)	37.7 (0.4%)
	Sub-total	1596 (70.3%)	6943 (76%)
REA (Off-grid)	Thermal	18 (0.9%)	35.1 (0.4%)
	Solar	0.569 (0.03%)	0.9 (−)
	Wind	0.55 (0.02%)	0.003 (−)
	Sub-total	19 (0.8%)	36 (0.4%)
IPPs	Thermal	516.82 (22.8%)	1188.9 (13%)
	Small hydro	0.814 (0.04%)	2.1 (0.02%)
	Geothermal	110 (4.9%)	955 (10.5%)
	Biomass	26 (1.2%)	14 (0.2%)
	Sub-total	654 (28.8%)	2160 (23.6%)
	Total	2269	9139

Electricity consumption by sector, 2014 (UNdata)

Kenya	Sector's consumption (GWh)
Total	7785.7
Food and tobacco	293 (3.8%)
Other industries	3695 (47.5%)
Households	1947.2 (25%)
Commercial and public services	1485.6 (19.1%)
Other	364.9 (4.7%)

Electricity consumption by customer type, 2015 (ERC)

Type of customer	Sales (GWh)
Domestic	1866 (26.3%)
Small commercial	1143 (16.1%)
Commercial and industrial	4030 (56.8%)
Off-peak (interruptible)	15 (0.2%)
Street lightning	35 (0.5%)
Total	7090

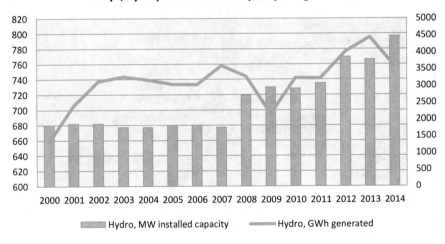

Hydropower installed capacity and generation, 2000–2014 (UNdata)

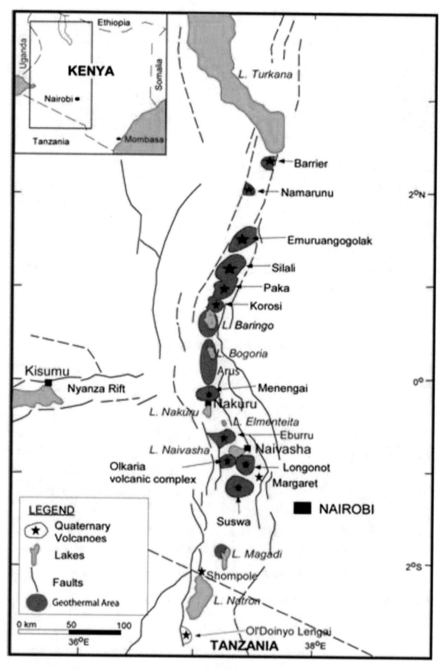

Geothermal potential in Kenya (Kenya Energy Regulatory Commission)

Malawi

Installed capacity and share of electricity generated, 2014 (UNdata)

Malawi	Installed capacity (MW)	Generation (GWh)
Total	501.45	2097.51
Hydro	290.35 (57.9%)	1916.5 (91.4%)
Thermal	211.1 (42.1%)	181.01 (8.6%)

Installed capacity as of ESCOM site table (2017)

Name	Capacity (MW)	Type
Kapichira Falls	129.6	Hydro
Nkula	124	Hydro
Tedzani	92.7	Hydro
Wovwe Mini Hydro	4.35	Hydro
Mzuzu Diesel Unit	1.1	Diesel
Likoma Islands Diesel Units	1.05	Diesel
Chizumulu Islands Diesel Units	0.3	Diesel
Total	353.1	

Electricity consumption by sector, 2014 (UNdata)

Malawi	Sector's consumption (GWh)
Total	1779
Manufacturing, construction and non-fuel industry	570 (32%)
Households	550 (30.9%)
Commercial and public services	230 (12.9%)
Other	429 (24.1%)

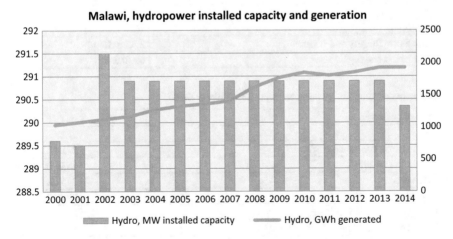

Hydropower installed capacity and generation, 2000–2014 (UNdata)

Mozambique

Installed capacity and share of electricity generated, 2014 (UNdata)

Mozambique	Installed Capacity (MW)	Generation (GWh)
Total	2682.4	17,739
Hydro	2322 (86.6%)	16,358.8 (92.2%)
Thermal	359 (13.4%)	1379 (7.8%)
Solar	1.36 (0.06%)	1.2 (0.01%)

Installed capacity available to Electricidade do Moçambique, 2015 (World Bank)

Project	Source	Installed capacity (MW)
Total		679[a]
HCB firm	Hydro	300
HCB non-firm	Hydro	200
Mavuzi	Hydro	52
Chicamba	Hydro	44
Corumana	Hydro	16
Pequenes Libombos	Hydro	2
Sub-total	Hydro	614 (90.4%)
Aggreko	Gas	15
Aggreko 2	Gas	32
Aggreko (Nacala)	Diesel	18
Sub-total	Thermal	65 (9.6%)

[a]This figure excludes the exports of electricity generated at Cahora Bassa dam to South Africa

Electricity consumption by sector, 2014 (UNdata)

Mozambique	Sector's consumption (GWh)
Total	12,342
Non-ferrous metals	8764 (71%)
Other industry	1887 (15.3%)
Households	1629 (13.2%)
Agriculture, forestry and fisheries	2.32 (0.02%)
Commercial and public services	60 (0.5%)

Electricity consumption by customer type, 2014 (World Bank)

Type of customer	Sales (GWh)
Total	3620
Transmission connected customers	371 (10.3%)
Residential customers	1536 (42.4%)
Commercial	345 (9.5%)
Agriculture	27 (0.7%)
Large customers low voltage	182 (5%)
Large customers medium/high voltage	1159 (32%)

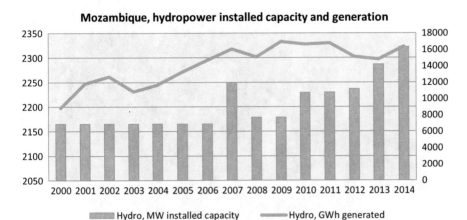

Hydropower installed capacity and generation, 2000–2014 (UNdata)

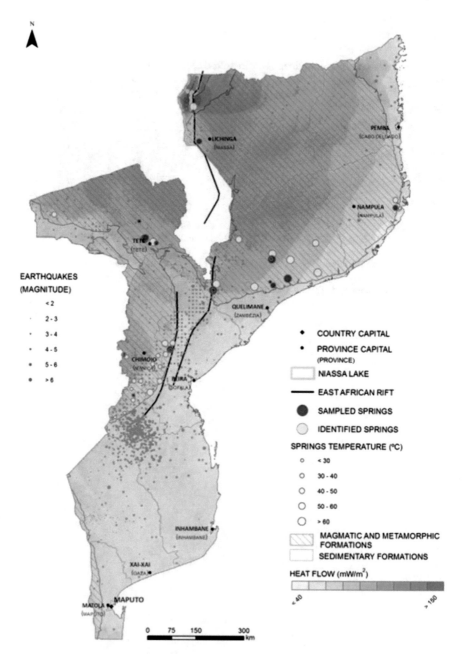

Geothermal potential in Mozambique (Energy Atlas Mozambique 2014)

Rwanda

Installed capacity and share of electricity generated, 2014 (UNdata)

Rwanda	Installed capacity (MW)	Generation (GWh)
Total	118.6	476.1
Hydro	76 (64.1%)	185.8 (39%)
Thermal	42 (35.4%)	290 (60.9%)
Solar	0.6 (0.5%)	0.3 (0.6%)

Evolution of installed capacity (MW) in Rwanda, 2010–2016 (Rwanda Energy Group)

Rwanda	2010	2011	2012	2013	2014	2015	2016
Total (MW)	84.9	90.3	100.3	104.1	140.6	170.6	174.6
Hydro	43.3	48.6	48.6	52.4	80.4	80.4	84.4
Diesel	37.8	37.8	47.8	47.8	47.8	51.8	51.8
Gas	3.6	3.6	3.6	3.6	3.6	29.6	29.6
Solar	0.3	0.3	0.3	0.3	8.8	8.8	8.8

Electricity consumption by sector, 2014 (UNdata)

Rwanda	Sector's consumption (GWh)
Total	438.3
Manufacturing, construction and non-fuel industry	82.9 (18.9%)
Households	355.4 (81.1%)

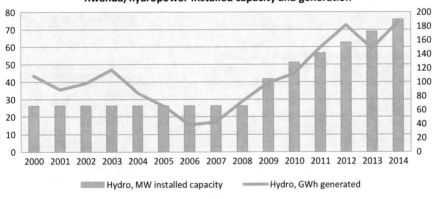

Hydropower installed capacity and generation, 2000–2014 (UNdata)

South Africa

Installed capacity and share of electricity generated, 2014 (UNdata)

South Africa	Installed capacity (MW)	Generation (GWh)
Total	46,963	252,578
Nuclear	1880 (4%)	13,794 (5.5%)
Hydro	725 (1.5%)	4082 (1.6%)
(of which pumped)	*411 (0.9%)*	*3107 (1.2%)*
Thermal	43,538 (92.7%)	232,512 (92.1%)
Wind	450 (1%)	1070 (0.4%)
Solar	370 (0.8%)	1120 (0.4%)

Eskom installed capacity as of February 2017 (Eskom)

Coal-fired stations	Nuclear station
Arnot: 2352 MW	Koeberg: 1940 MW
Camden: 1561 MW	**Conventional hydro stations**
Duvha: 3600 MW	Gariep: 360 MW
Grootvlei: 1180 MW	Vanderkloof: 240 MW
Hendrina: 1893 MW	**Pumped storage schemes**
Kendal: 4116 MW	Drakensberg: 1000 MW
Komati: 990 MW	Palmiet: 400 MW
Kriel: 3000 MW	Ingula: 1332 MW
Lethabo: 3708 MW	**Gas-fired stations**
Majuba: 4110 MW	Acacia: 171 MW
Matimba: 3990 MW	Port Rex: 171 MW
Matla: 3600 MW	Ankerlig: 1338 MW
Medupi: 794 MW (Unit 6)	Gourikwa: 746 MW
Tutuka: 3654 MW	**Windfarm:** Sere: 100 MW

Electricity consumption by sector, 2014 (UNdata)

South Africa	Sector's consumption (GWh)
Total	198,093
Iron and steel	3698 (1.9%)
Chemical and petrochemical	11,314 (5.7%)
Non-ferrous metals	16,797 (8.5%)
Mining and quarrying	30,609 (15.5%)
Other manufacturing, constr. and non-fuel ind.	57,842 (29.2%)
Transport (Rail)	3172 (1.6%)
Transport (Other)	601 (0.3%)
Households	37,777 (19.1%)
Agriculture, forestry, fishing	5562 (2.81%)
Commercial and public services	27,455 (13.9%)
Other	3266 (1.7%)

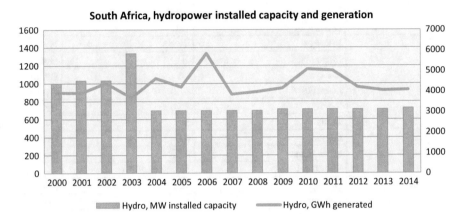

Hydropower installed capacity and generation, 2000–2014 (UNdata)

Tanzania

Installed capacity and share of electricity generated, 2014 (UNdata)

Tanzania	Installed capacity (MW)	Generation (GWh)
Total	1115	6219
Hydro	561.8 (50.4%)	3611 (58.1%)
Thermal	546.2 (48.99%)	2590 (41.7%)
Solar	7 (0.63%)	18 (0.3%)

Installed capacity, 2014 (Ministry of Energy and Mineral)

Fuel	Installed capacity (MW)
Total	1583
Hydro	561 (35.4%)
Natural gas	527 (33.3%)
HFO[a]/GO[b]/Diesel	495 (31.3%)

[a]Heavy fuel oil
[b]Gasoil

Installed capacity, 2016 (Ministry of Energy and Mineral)

Owner	Name	Capacity (MW)	Fuel
Tanesco	Hale	21	Hydro
	Nyumba Ya Mungu	8	Hydro
	Pangani Falls	68	Hydro
	Kidatu	204	Hydro
	Mtera	80	Hydro

(continued)

Owner	Name	Capacity (MW)	Fuel
	Uwemba	0.843	Hydro
	Kihansi	180	Hydro
IPPs	Mwenga	4	Hydro
	Mapembasi	10	Hydro
	EA Power	10	Hydro
	Darakuta	0.46	Hydro
	Yovi	0.96	Hydro
	Tulila	5	Hydro
	Ikondo	0.6	Hydro
	Mbangamao	0.5	Hydro
	Sub-Total	593.4 (42.69%)	
Tanesco	Ubungo 1	102	Gas
	Tegeta GT	45	Gas
	Ubungu 2	105	Gas
	Zuzu D	7	IDO[a]
	Nyakato	63	HFO[b]
	Kinyerezi	150	Gas
IPPs	Songas 1	42	Gas
	Songas 2	120	Gas
	Songas 3	40	Gas
	Tegeta IPTL	103	HFO
	TPC	17	Biomass
	TANWAT	2.7	Biomass
	Sub-Total	796.7 (57.31%)	
	Total	1390	

[a]Industrial diesel oil
[b]Heavy fuel oil

Electricity consumption by sector, 2014 (UNdata)

Tanzania	Sector's consumption (GWh)
Total	4976
Manufacturing, construction and non-fuel industry	1270 (25.5%)
Households	2227 (44.8%)
Commercial and public services	1141 (22.9%)
Agriculture, forestry and fishing	180 (3.6%)
Other	158 (3.2%)

Electricity consumption by sector, 2015 (Ministry of Energy and Mineral)

Sector	Consumption (GWh)
Total	6320
Industry	2410 (38.1%)

(continued)

Sector	Consumption (GWh)
Commercial and services	300 (4.7%)
Zanzibar	340 (5.4%)
Gold	200 (3.2%)
Residential	1990 (31.5%)
T/D loss	1080 (17.1%)

Gas and coal plants under development, 2016 (Ministry of Energy and Mineral)

Name	Capacity (MW)	Fuel
Kynierezi I–IV	1355	Gas
Somanga/Somanga Fungu	860	Gas
Bagamoyo	200	Gas
Mtwara	318	Gas
Mchuchuma I–IV	600	Coal
Kwira I–II	400	Coal
Ngaka I–II	400	Coal

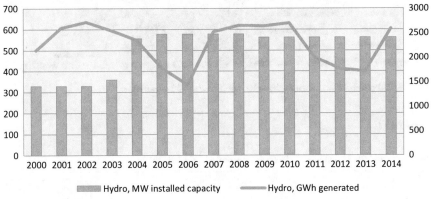

Hydropower installed capacity and generation, 2000–2014 (UNdata)

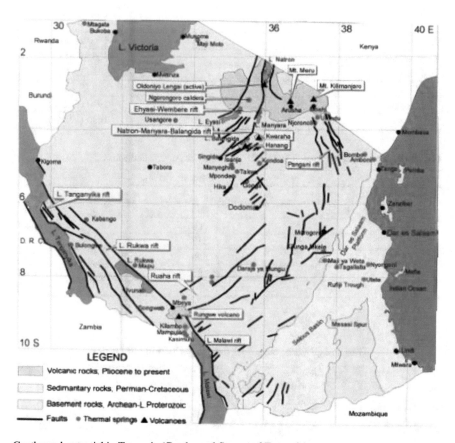

Geothermal potential in Tanzania (Geothermal Survey of Tanzania)

Uganda

Installed capacity and share of electricity generated, 2014 (UNdata)

Uganda	Installed capacity (MW)	Generation (GWh)
Total	883.3	3257.7
Hydro	705 (79.8%)	697.1 (21.4%)
Thermal	178.3 (20.2%)	2560.6 (78.6%)

Installed capacity and share of generated electricity, 2017 (Electricity Regulatory Authority)

Uganda	Installed capacity (MW)	Generation (GWh)
Total	947	3856
Hydro	700 (73.9%)	3441 (89.2%)
Thermal	140 (14.8%)	242 (6.3%)
Biomass	96 (10.1%)	150 (3.9%)
Solar	11 (1.2%)	23 (0.6%)

Electricity consumption by sector, 2014 (UNdata)

Uganda	Sector's consumption (GWh)
Total	2416.9
Iron and steel	1068.8 (44.2%)
Other industries	533.2 (22.1%)
Households	519.4 (21.5%)
Commercial and public services	287.5 (12%)
Other	8 (0.3%)

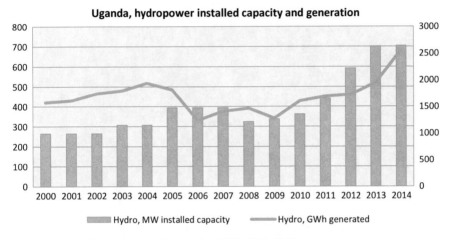

Hydropower installed capacity and generation, 2000–2014 (UNdata)

Appendix B: Methodology and Parameters of the Electrification Analysis

Input datasets and sources

Dataset	Description	Source
Administrative boundaries	National administrative boundaries to define the spatial extent and crop other datasets	Database of Global Administrative Areas (2018)
Digital elevation	Elevation (in meters)	NASA LP DAAC (2011)
Small hydropower potential	Position, potential (MW)	Korkovelos et al. (2017)
Land cover	Categories of predominant land-cover define land suitability for installing different generation technologies	Channan et al. (2014)
Night-time lights	Employed to calibrate the population without access	NASA (2017)
Population	Number and position of the population within national boundaries	WorldPop/Linard et al. (2012)
Roads	Employed to calibrate the population without access	CIESIN and ITOS (2013)
Slope	Calculated from DEM datasets	Authors' elaboration
Solar PV potential	Global horizontal irradiation to calculate solar PV potential	SolarGIS (2017)
Solar restrictions	Calculated from land cover dataset to restrict PV installation in certain land settings (e.g. cropland and water bodies)	Authors' elaboration
Electricity substations	Employed to calibrate the population without access	Energydata.info (2018)
Current and planned electricity transmission network	Employed to define the cost and potential for new connection and grid extension	Arderne/World Bank (2017)

(continued)

Dataset	Description	Source
Travel time to the nearest 50,000+ city	Defined to calculate the LCOE of diesel	Weiss et al. (2018)
Wind potential	In m/s, used to calculate the wind power capacity factor	DTU/Global Wind Atlas (2017)

Common user-defined parameters

Parameter	Description	Value
Discount rate	To weight results from the present generation's perspective (relatively less importance is given to the future)	10%[a]
MV line cost (USD/km)	1–66 kV	6000 USD
LV line cost (USD/km)	<1 kV	3000 USD
HV line cost (USD/km)	>66 kV	30,000 USD
HV to LV transformer cost (USD/unit)	Cost of a transformer between transmission and distribution grid	4000 USD
Grid connection cost per household	The average charge to be borne by a household (unless a subsidy policy is in place) to get grid electricity at home	450 USD
Operation and maintenance costs of transmission and distribution lines as % of capital costs	Share of O&M costs over the total capital costs to be borne by the electricity supply company for grid O&M	5%
Grid capacity investment (USD/kW of on-grid added capacity excluding the grid itself)	The public or private average investment required to add new capacity to the national grid-connected electricity generation	Country-level, depending also on generation mix considered. Average of 2000 USD/KW
Diesel gen-set mini grid investment cost	Average unit (per kW) cost of installing, operating and maintaining the system	721 USD/kW + 10% O&M costs (% of investment cost/ year)
Small hydro mini grid plant	Average unit (per kW) cost of installing, operating and maintaining the system	5000 USD/kW + 2% O&M costs (% of investment cost/ year)
Solar PV mini grid	Average unit (per kW) cost of installing, operating and maintaining the system	3200 USD/kW + 1.5% O&M costs (% of investment cost/year)
Wind turbines mini grid	Average unit (per kW) cost of installing, operating and maintaining the system	3000 USD/kW + 2% O&M costs (% of investment cost/ year)
Diesel standalone investment cost	Average unit (per kW) cost of installing, operating and maintaining the system	938 USD/kW + 10% O&M costs (% of investment cost/ year)

Pueyo et al. (2016)

Country-specific parameters

Country	Pop (mil.)	Urban pop (%)	Pop 2030 (mil.)	Urban pop. 2030 (%)	People per HH, rural	People per HH, urban	Grid capacity inv. cost (USD/kW)	Grid losses (%)	Electr. rate (%)
Burundi	10.20	12.06	17.36	35	5	3.5	2200	17.65	8
Kenya	47.24	25.62	65.41	35	4.5	5	1700	17.50	64.5
Malawi	17.57	16.27	26.58	30	6	4	2000	17.65	11.3
Mozambique	28.01	32.21	41.44	50	5.5	4	2000	14.70	28.6
Rwanda	11.63	28.81	15.78	40	4.5	3.5	2000	17.65	30.0
Tanzania	53.88	31.61	82.93	50	5.5	4	1700	17.65	32.7
Uganda	40.14	16.10	61.93	35	5	3.5	2000	17.65	19.4

All the remaining technical and economic parameters (e.g. technology investment costs, efficiency factors) for each technology were left as default in the model (refer to Mentis et al. 2017, to the OnSSET documentation, and to the model code)

Maps of least-cost technology

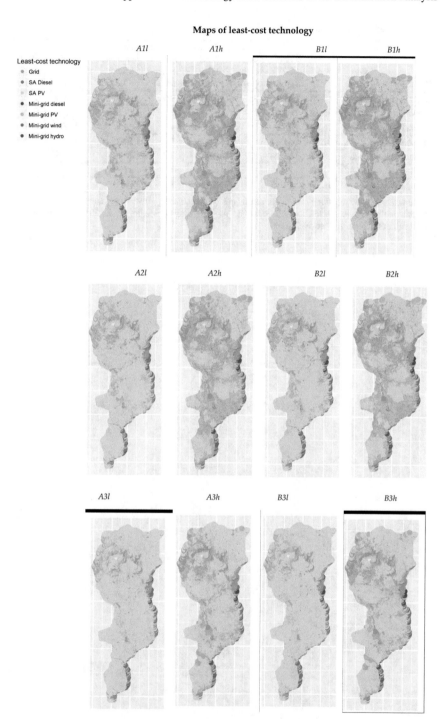

Maps of least-cost technology

References

Center for International Earth Science Information Network—CIESIN—Columbia University, Information Technology Outreach Services—ITOS—University of Georgia (2013) Global Roads Open Access Data Set, Version 1 (gROADSv1). NASA Socioeconomic Data and Applications Center (SEDAC), Palisades, NY

Channan S, Collins K, Emanuel WR (2014) Global mosaics of the standard MODIS land cover type data. University of Maryland and the Pacific Northwest National Laboratory, College Park, MD

DTU Technical University of Denmark (2017) Global wind atlas. Retrieved from http://globalsolaratlas.info/api/download

Energydata.info (2018) Energydata.info. https://energydata.info

Korkovelos A, Mentis D, Hussain Siyal S, et al (2017) A geospatial assessment of mini/small hydropower potential in Sub-Saharan Africa, p 6825

Linard C, Gilbert M, Snow RW, et al (2012) Population distribution, settlement patterns and accessibility across Africa in 2010. PLoS One 7:e31743. doi: https://doi.org/10.1371/journal.pone.0031743

Mentis D, Howells M, Rogner H et al (2017) Lighting the world: the first application of an open source, spatial electrification tool (OnSSET) on Sub-Saharan Africa. Environ Res Lett 12:085003. https://doi.org/10.1088/1748-9326/aa7b29

NASA LP DAAC (2011) ASTER global digital elevation map v2. https://asterweb.jpl.nasa.gov/gdem.asp. Accessed 6 Aug 2018

Pueyo A, Bawakyillenuo S, Osiolo H (2016) Cost and returns of renewable energy in Sub-Saharan Africa: A comparison of Kenya and Ghana. IDS

SolarGIS (2017) Potential solar PV output raster files. Retrieved from https://solargis.com/maps-and-gis-data/download/

The World Bank (2017) World Bank Data. Accessed 20 Nov 2017

Weiss DJ, Nelson A, Gibson HS et al (2018) A global map of travel time to cities to assess inequalities in accessibility in 2015. Nature 553:333–336. https://doi.org/10.1038/nature25181

Printed in the United States
By Bookmasters